RURAL PROPERTY PLANNING

Farm Buildings

PLANNING AND CONSTRUCTION

Neil Southorn

INKATA PRESS

INKATA PRESS

A division of Butterworth-Heinemann Australia

Australia
Butterworth-Heinemann, 18 Salmon Street, Port Melbourne, 3207

Singapore
Butterworth-Heinemann Asia

United Kingdom
Butterworth-Heinemann Ltd, Oxford

USA
Butterworth-Heinemann, Newton

National Library of Australia Cataloguing-in-Publication entry

Southorn, Neil.
 Farm buildings : planning and construction.

 Includes index.
 ISBN 0 7506 8934 X.

 1. Farm buildings - Design and consruction. I. Title.
 (Series : Practical farming).

728.92

©1996 Neil Southorn

Published by Reed International Books Australia. Under the *Copyright Act* 1968 (Cth), no part of this publication may be reproduced by any process, electronic or otherwise, without the specific written permission of the copyright owner.

Enquiries should be addressed to the publishers.

Typeset by Ian MacArthur, Hornsby Heights and Belbora, NSW.
Printed in Australia by Ligare Pty Ltd, Riverwood, NSW.

Contents

Introduction   7
Safety   9

CHAPTER 1: *General planning objectives*   11
   Site   11
   Function and layout   18
   Other building features   21
   Procedural matters   22

CHAPTER 2: *Construction materials*   24
   Steel   24
   Concrete   28
   Timber   38
   Fasteners   46

CHAPTER 3: *Site measurements and plans*   52
   General measurements   52
   Measurement of distances   55
   Measurement of angles   60
   Measurement of elevation   60
   Standard plan presentation   65
   Pegging   66

CHAPTER 4: *Aspects of structural design*   71
   Load estimation   71
   Design of structural components   76

CHAPTER 5: *Types of construction*   80
   Portal frame construction   80
   Pole-in-ground construction   92
   Arch construction   95
   Concrete slab construction   97
   Raised floors   99

CHAPTER 6: *Cladding and building protection* — **102**
  Cladding — 102
  Weatherproofing and roof drainage — 110
  Protection from pests — 112

CHAPTER 7: *Building services* — **115**
  Electricity — 115
  Water — 116
  Access — 117
  Ventilation — 118

CHAPTER 8: *Inspection and maintenance* — **120**
  Inspections — 120
  Maintenence and protective measures — 122

CHAPTER 9: *Special purpose buildings* — **126**
  Greenhouses — 126
  Shearing sheds — 131
  Grain sheds and silos — 137
  Intensive pig housing — 138
  Other specialised buildings — 142

References — 147

Acknowledgments — 148

Index — 149

Introduction

Most rural properties have a large number of buildings which contribute to the smooth and efficient running of the farm. At least, that is how it should be. Buildings which are poorly located, inadequately maintained or just badly designed do not help much at all. A badly designed shearing shed, for example, will require extra labour during shearing, and could reduce the quality of the job that is done, resulting in downgrading of the wool that is being harvested. Good machinery storage and a well planned workshop will help maintain the value of plant and equipment, and reduce wasted time during critical periods. Consequently, this book discusses the various factors to consider in regard to the part buildings play in the overall operations of the farm.

Many farmers have the skills necessary to fabricate and erect farm buildings. For those who do not, this book provides a basic introduction to the materials used. (Specific fabrication skills, such as cutting and welding, are covered in other books in this series.) Even so, a decision has to be made whether to proceed with construction yourself, to purchase a kit shed, or to call in contractors to do the job, and information to help make that decision is presented.

The book describes the general principles behind the structural design of farm buildings, so that the various features of a structure can be identified and considered. It will also help when repair jobs are required. Bear in mind that the structural design of farm buildings is a complex task, requiring certification by appropriately qualified structural engineers, and is well outside the scope of this book.

The various parts that make up a building are described. From a structural viewpoint, each part has an important role to play in ensuring the integrity of the structure: that it serves the purpose it was designed for, and stands up against the elements. Some of the more important buildings on farms, both general agricultural and horticultural, are singled out for specific discussions, as are their maintenance requirements.

When planning a new building, always go and look at sites where the same or a similar building has been constructed, or built by the contractor you are considering engaging. This will help ensure that what you get is what you want, not just from the quality of the work in construction, but also that the functional design of the building is adequate.

More specific sources of information should also be consulted. These include:

- building contractors, experienced in rural applications

- various building codes
- building inspectors at the local council
- specialists or consultants in the building industry; architects, structural engineers, general building consultants, etc
- for special purpose rural buildings, advisory officers in government agricultural departments.

It might seem that the emphasis is on new buildings. However, an understanding of construction principles and practices is equally applicable to maintenance and repair jobs.

Safety

The construction industry has a high incidence of workplace injury. Although farm buildings are not the same type of project as city skyscrapers, injury is still easily possible, particularly when the poor accident record of rural industries is taken into consideration.

- Lifting and manual handling on a construction site can easily cause back injury. Make lifting equipment available to minimise the amount of manual handling, and plan the structure with ease of erection in mind.
- Take care during excavation and levelling to avoid the normal risks of machinery operations and trenching.
- Wear appropriate personal protection equipment. This includes eye and ear protection as required, but also a hard hat to reduce risk of injury from falling objects, and safety boots. Use gloves to protect the hands. Use sunscreen protection on bare skin, and sunglasses against reflected radiation from cladding and insulation.
- Use scaffolding in preference to ladders. Make sure the scaffolding is erected

Use scaffolding in preference to ladders when working high up, but make sure it is installed on firm ground

according to the required standards, and located on firm ground. Fixed and mobile scaffolding can be hired, as can cherry pickers and scissor lift work platforms.
- If ladders are used, they should be secured against the building frame, and the person secured onto the ladder, using an appropriate harness. In some cases, it may be necessary to fix a device to the frame of the building specifically for securing a ladder. Ladders should be sound, with non-slip feet, and be located at the correct incline. When climbing up to a roof, extend the ladder one metre past the roof line. There is a limit to the height possible with extension ladders. Do not stand past the third last rung on any ladder (or as directed by the ladder manufacturer).
- When working on rooves, special precautions are required, for obvious reasons. A wire mesh, of particular specifications and method of fixing and joining, must be installed above the roof purlins, prior to installation of cladding (and insulation). This is to protect persons who might fall off or through the roof. The installation of the mesh must follow safe work practices. If mesh is not used, some other type of "fall arrest system", guard rail or safety net should be in place. Refer to *Code of Practice — Safe Work on Roofs* for further details (available from NSW Workcover).

CHAPTER 1

General planning objectives

A number of factors needs to be considered when constructing new farm buildings before any plans or specifications are drawn up. Buildings should be located in the best position to maximise their benefit to farm operations, and to help ensure their long life. The ground at the site (foundation) must be sound, and of sufficient strength to support the structures. Even the direction they face (aspect) can sometimes be important.

Having decided the location, the building's shape, size and internal environment must also suit the purpose it is designed for, and some people insist on a building having a particular appearance, or being made of particular materials. Frequently, details of the building must be approved by various authorities.

Once these general planning decisions have been made, plans and specifications can be considered, and costs determined.

Site

The best location for a building is determined by a number of factors. Some of these affect the structural integrity of the building, while others relate more to its function.

Foundation suitability

The foundation of the building is the natural ground material the structure rests upon. The footings are the below ground part of the structure that transfer the load of the building to the foundation. The foundation must be sufficiently sound to prevent movement of the building.

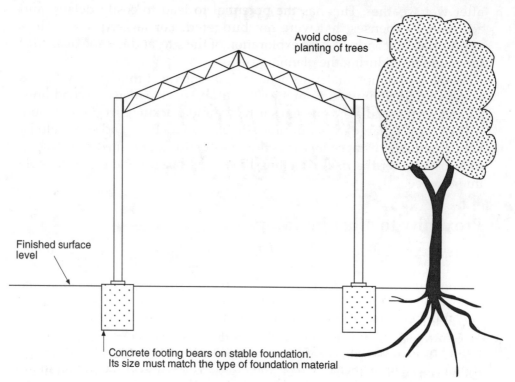

Figure 1.1 *Footing dimensions must match foundation conditions. Avoid trees*

Some ground conditions are more suitable than others for a building's foundation. The worst materials are soft swelling clays that crack when they dry, loose unconfined sand, and floating rocks that move about in the soil. Apart from these extreme cases, it should be possible to design footings which give a sound base for the building. However, each foundation material will have different strength characteristics, which, if exceeded, will allow the structure to sink into the ground. This is bad enough, but if part of the structure sinks and another part doesn't, then the structure may crack and become unsafe. To prevent subsidence the base of the footing needs to be of sufficient size that the downward force on the footing is spread over an adequate area of the ground. The bearing strength of the foundation material will then not be exceeded, and the building will be secure.

Where a building must be built on a foundation which is less than ideal, special construction methods must be employed for the footings. A number of techniques are available; for example, sinking piers to more stable foundation material at greater depth, or the use of reinforced beam concrete slabs so the structure "floats". Such methods add to the cost of the structure. In these cases specialist advice is necessary to test the strength and stability of the foundation, and to design the footings.

Occasionally the site may appear satisfactory, but unstable material is uncovered during levelling or excavation for the footings, or is encountered after wet weather. This has the potential to lead to costly delays, and additional expenses that were not budgeted. For larger projects, it is appropriate to conduct some exploration of the site, and test the quality of the foundation during the planning stages.

Some foundation materials behave differently wet to when dry. Excess water should be drained from the site, but reasonable moisture conditions should be maintained to prevent shrinkage and cracking of the foundation.

The roots of large trees can cause damage to footings and so should be kept a suitable distance away. Buildings should not be sited too close to trees because of the risks of falling limbs, fire hazard and other possible interference.

Proximity to other buildings

There are several advantages in locating farm buildings near each other. The principal gain is that it provides a centralised operations area, where machinery, workshop, fuel store and so on are all close together. On some properties this has the potential to save time in repairs and maintenance, and at the start of each day's work or shift.

Centralisation can also reduce the overall costs of building development, since different buildings can share the cost of services. In the case of electricity connection, a single transformer can serve a number of buildings, with low voltage power (usually 240 or 415 Volts) distributed by overhead lines, or underground in conduit if the distance is only a few hundred metres. A large part of the cost of connecting electricity to a site is in the transformer at the end of the high voltage line, so needing only one is a saving.

Savings can similarly be made if drinking water and amenities are located at a convenient central location, and rainwater from the roof areas of adjacent buildings can be collected in a single large water tank.

There are some possible exceptions to the idea of grouping buildings together. For example, hay sheds are better located away from the main working areas of the farm because of the fire risk of stacked hay. Many farms have hay sheds in or near the hay paddock, to minimise the time necessary to get the hay under cover, and also it is closer to livestock. The same applies to grain storage.

Fuel should be stored a safe distance from other facilities, particularly from workshops where welding and grinding occur.

The location of the homestead in relation to the other farm buildings may also need special consideration. It may be located nearby, for convenience in getting to work, or because of the cost of separate electricity and water supplies. However, there is a risk that noise, dust and smell associated with normal farm activities will upset family members. If farm buildings are to be located some distance away, requiring vehicle transport, it may as well be a reasonable distance, because the extra travel time will not make a great deal of difference.

Figure 1.2 *This shed has been destroyed by fire. Moist hay can spontaneously combust, so keep hay sheds separate from other buildings on the property*

It is possible to locate buildings too close together. There may be insufficient space when repairs or maintenance are required for example. Although having electricity nearby is normally necessary, overhead conductors can be a significant safety hazard when planning buildings. Consider the long term future, and leave room for expansion. It could be advantageous to use a construction method that is easy to add on to.

Access

Many farm buildings require access by heavy vehicles. A single gravelled road leading to a complex of buildings will be the most cost effective, particularly if they are located near the property boundary to reduce the distance required for heavy duty roads.

Some buildings require access by semi-trailers for cartage of hay, grain, horticultural produce, livestock and wool, and possibly general freight. Even if semi-trailers are not normally used, it is best to at least plan for such vehicles in the future. Consequently a substantial road surface and sufficient area for manoeuvring need to be planned. This also requires a stable foundation.

A road width of 4 metres is considered a minimum, with a compacted gravel surface of 3.5 metres preferred. Gravel depth should be around 100 millimetres, depending on the quality of the road base used. It should be

Figure 1.3 *An example of poor site planning — the overhead powerlines at this site prevent tipping semi-trailers unloading at the silos, and represent a major safety hazard when moving augers. Note the crumpled side of the silo, crushed by an auger, which could weaken the structure*

formed up higher than the surrounding ground level, and fully culverted to provide adequate drainage. Avoid putting such a road over difficult sites and keep clear of overhanging timber. Because such a road is expensive, there is some incentive in keeping it as short as possible. When planning new buildings that require heavy vehicle access, the cost of the access may influence the location of the buildings.

Slope

The construction of most farm buildings requires a level site. However a perfectly level site will be poorly drained. In such cases ground level under the building is raised to facilitate removal of stormwater run-off. This is achieved by pushing earth toward the building, or by bringing in material from elsewhere. Make the area larger than the floor area of the building itself, graded slightly for run-off. The footings supporting the building must penetrate this fill into undisturbed material below.

Where the building is to be located on a moderately sloping site, it is normal practice to cut and fill the area to provide a level surface, again for an area larger than the floor area of the building. The steeper the site, the greater the depth of cut and fill. This adds to the cost of preparing the site, and possibly requires specially designed footings, both of which are extra to the cost of the building itself. Where footings have to be located in the fill

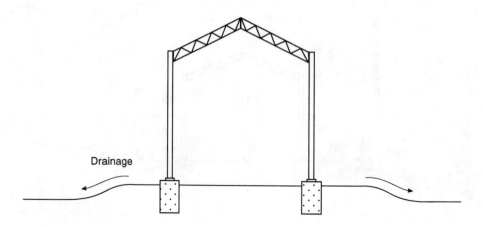

Figure 1.4 *On a level site the building should be raised with fill to provide drainage. Note the footings must extend to undisturbed foundation material*

Figure 1.5 *This may be a perfectly good shed, but it is difficult to get to in wet weather because of poor drainage*

General planning objectives

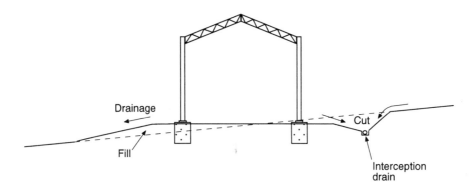

Figure 1.6 *Specially designed footings may be necessary on steeper sites and site drainage may be more complicated*

area, they must be dug into undisturbed material, using deep piers if necessary.

Where sloping ground is above floor level, there is a risk run-off water will enter the building. Ensure sufficient fall away from the building perimeter, and install sufficient drainage capacity to intercept or contain heavy storm run-off, including roof drainage. Contour banks or ditches may also be necessary to divert run-off.

It may be necessary to split the floor level of the building on steeper slopes. Occasionally, the slope can be used to advantage in a raised floor shed, such as a shearing shed, to provide extra storage space and headroom below the floor. However, both these options require a non-standard design of the wall columns since their height is not uniform throughout the structure. This can add a penalty to the cost, particularly for kit sheds, and it may be preferable to choose a more level site to start with.

Aspect

Aspect refers to the orientation of the building, that is the direction it faces.

General purpose buildings, such as machinery sheds, that have one side open, are generally aligned with the open side facing north. This allows the winter sun to penetrate into the building, but limits summer sun by the roof line. An eave would help provide summer shade, but this is not normally provided on general purpose farm buildings. This orientation may also help shelter the building from south-westerly winter winds, typical in south-east Australia. Local information should be applied to each site. For example local terrain may influence wind patterns, such as wind tunnel effects caused by hills and gullies, or the site may be subject to troublesome winds from other directions. Shelter may be better provided by a treelot.

In the case of a shearing shed, its aspect may be governed by its relationship to the adjoining sheepyards. The preferred orientation of the building needs to be considered in conjunction with the yards. For example, the sheepyard race may be best running north, but it should also run away from the shed, to encourage sheep to run up the race. This may dictate the position and orientation of the shed.

Some structures require much more detailed consideration of aspect. Greenhouses rely on solar radiation penetrating the structure, and its orientation can control this factor. A greenhouse aligned with its long axis east-west will have more wall and roof area exposed to winter sun compared to a north-south alignment, which will maximise sunlight and heat input in winter. It is also important to consider the shading caused from greenhouse frame components. It is preferred that they move through the day, rather than casting a permanent shadow on plants inside. These factors would be more important in southern areas (further south than say 35 degrees) than nearer the equator, where lack of winter sun is less of a problem than reducing summer heat.

Similar considerations apply to housed livestock, such as pigs. Orientation of their buildings can influence the solar heat load, a problem to be removed by good ventilation in summer. An east-west alignment helps here by minimising building area exposed to summer sun.

Function and layout

Dimensions

The overall floor size will determine length and width of the building, which in turn determine the span of the frame and the width between frames.

For standard non-domestic type constructions, these dimensions are limited to a range of standard sizes; typically a 6 to 18 metre span, and 4.5, 6 or 7.5 metres between spans to form a bay, and multiple bays to give the required length of building. Consequently most farm buildings are long and rectangular in shape rather than square. Curves in a floor plan can be built, but are usually expensive. As the span and width between columns increases, then the components of the frame must be stronger, and therefore heavier, to carry the load. This makes them more expensive per square metre of floor space. The same applies to the height of the columns; many standard heights are available, but there is no need to go higher than necessary. Some particular height may be required to get specific vehicles into the building, or for loading produce.

The overall volume of air contained within the building plays an important part in determining the ventilation requirements if livestock are housed in the building or if it is a coolstore or a greenhouse, or merely for worker comfort.

Suspended loads

Standard building construction does not usually cater for any loads carried by the frame of the building, apart from the weight of the building components themselves. Suspended loads, which can range from a steel rack in a workshop to bulk material storage or heavy duty lifting hoist, may need to be considered in the planning stages.

Environment control

It may be required that the internal environment of the building can be controllable to some extent. These factors are best addressed at the planning stage.

Insulation can help control temperature. Foil is commonly used as a reflective insulation, while fibreglass batts, polyurethane or polystyrene are used as mass insulation to control both high and low temperatures, particularly in the roof. If insulation is to be installed it must be supported in some way. If the building does not have a ceiling, then wire netting may be necessary.

Ventilation can be quite useful to control excessive summer temperatures, and the correct placement of vents needs to be considered. If possible and necessary, cross-flow ventilation is best, combined with a ridge vent. By placing a vent in the ridge of the roof, hot air will rise and escape from the building even in still air. Fans can be used in special buildings like greenhouses and pig housing, to force the ventilation rate up to the required number of air changes per hour.

Skylights can be useful in supplementing artificial light, or may be used alone in low cost buildings. Care needs to be taken to avoid bright spots and shadows within the building and excessive heat build up. Adequate lighting can often be obtained by side windows or, failing this, the use of fluorescent lighting is warranted in such buildings as farm workshops or woolsheds.

Interior partitions

If the floor plan of the building requires separate rooms with partition walls for them, then it needs to be decided at the planning stage whether they will be load bearing walls or not. Generally speaking, it simplifies construction if they are not, but they may need to be if a loft, mezzanine or upper floor is to be built, or if the roof profile is an unusual one. If internal walls are load bearing, they and their footings must be designed with these loads in mind.

Figure 1.7 *Translucent sheeting installed in the roof of this packing shed allows natural light, but can also increase summer temperatures*

Special purpose buildings

Buildings housing livestock, including shearing sheds, and buildings where frequent movement of produce and personnel are involved will be more efficient workplaces if sufficient planning goes into their proper layout. For example, a well designed shearing shed will hold enough sheep, will be easy to move and pen sheep, will be pleasant to work in, and will be easy to load wool from. It will enable greater control over woolclassing and subsequent clip preparation. Consequently, it can save labour costs during shearing, possibly increase shearer tallies, and even increase returns on the clip. These factors are associated with the layout of the shed, in addition to its structural design.

As a further example, pig production is greatly influenced by environmental conditions, which are largely associated with building design and functional layout. These special purposes must be addressed at the initial planning stages. Further details on some aspects of special purpose buildings are contained in Chapter 9.

Effluent disposal

Large volumes of waste are generated in buildings housing poultry or pigs in particular. In the case of poultry, it is mostly of low moisture content, able to be kept in the building over a period of time, and removed occasionally with a change of litter for broilers, or as the need arises with egg layers.

In the case of intensively housed pigs, large volumes of water are also involved, the liquid effluent requiring containment, treatment and disposal within the property.

In these situations, the structure must be designed with waste handling in mind, and using construction materials that can tolerate the corrosive environment that is generated. Even the dung collecting under a shearing shed floor can cause rapid deterioration of steel and timber supporting structures, particularly when combined with moisture. Storm run-off from cattle feedlots, although not a building as such, is also contaminated and must be collected, treated and disposed of accordingly.

The design of efficient management systems will be subject to scrutiny by the relevant environment protection authorities.

Other building features

Appearance

Some people place greater importance on appearance than others. Sometimes local council regulations will require a certain standard of appearance, to minimise the visual impact of the building. This usually refers to the colour of the external cladding, the height of the building, or its location in the landscape. The appearance of a building can be influenced by its general shape, such as the ratio of its height to length and width. The method of construction could be selected to provide a particular external appearance, if required.

Fire planning

A few basic rules can be followed to help reduce the risk of major fire damage to farm buildings:
- Reduce the amount of fuel in proximity to buildings by creating wide firebreaks preferably to bare earth or slashed green grass. Clear gutters and immediate area of leaves and other debris.
- The use of tree lots to reduce wind speed, and therefore the rate of progress of a fire, should not be overlooked, but keep trees a safe distance away from buildings.
- Buildings should be designed to prevent the entry of burning embers during a fire. Wall sheeting should be continuous from ground to eave with no timber parts of the structure protruding. Eaves should be boxed in to prevent embers being blown into the roof.
- Domestic dwellings require a reserve of water for fire fighting purposes, and this concept can be extended to other farm buildings.
- Locating farm buildings in close proximity can assist by reducing the length of firebreaks and it becomes easier to install sprinkler protection.

Procedural matters

Development application

A number of steps are required before construction of any major structure can commence. The first of these is the lodgement of a Development Application with the local council. This consists of a general description of the development proposal, including site plans, type and size of buildings, purpose of the buildings, anticipated vehicle movement, environmental impact, method of drainage etc. This does not usually require detailed specifications, but sufficient information to enable the council to decide whether the project should go ahead, and if so, what special conditions such as drainage requirements or limitations to building size or appearance, might apply. Most farm buildings present no problems, and some shires exempt standard farm buildings from this procedure.

Stricter requirements would be required if the development is a "prescribed" one. In agriculture, intensive development projects such as piggeries, feedlots, abattoirs, and other large processing operations would fall into this category. If this is the case, council would require a much more substantial submission, including assessment of environmental impact, and compliance with relevant regulatory authorities.

Building application

Once the development application is approved, then a Building Application is submitted. These documents contain full details and specifications of all parts of the project, including structural specifications, plans, and cross-sections of all buildings and other works. This is to ensure that all specifications are satisfactory and meet the appropriate building codes. If the council is in any doubt, a structural engineer's certificate may be required to verify adequacy of the specifications. The Building Application therefore requires all construction details to be finalised, including the site levels, although minor changes can be made during construction if circumstances require them. The manufacturers of kit buildings supply plans ready to submit to councils. Details of electricity and water connections will also be required, if applicable.

Planning and construction phase

Once plans have been approved, you are in a position to call for quotes or tenders. Some organisations require a formal process for this aspect of planning, to ensure equity between bidders, and to obtain the best price that meets specifications. On large projects, seek specialised assistance for the preparation and evaluation of tender documents and contracts, as well as the estimation, ordering and delivery of materials, and the supervision and payment of contractors. The success and timing of these duties will play an important part in the overall success of the project.

Even for small projects where a formal process is not used, consider the following points to be important features:

- Use approved plans and specifications as the basis for costing and construction, even if it is agreed they be varied. Be involved in the important decisions as they are made, and ensure the work goes according to the plans and specifications.
- Before construction contractors move onto the site, some site preparation will be necessary. All-weather access is preferred, and you may need electricity and water to be available to run power tools and mix concrete. A generator and temporary water tank may be necessary.
- Make sure that contractors have all necessary licenses, worker's compensation insurance and so on.
- Do not guess where existing telephone, water or electricity lines might be located. Contact the local authorities and have them mark their location on the ground.

Progress reports

During construction you may be obliged to alert the council that various stages of construction are ready for inspection. These are usually:

- the excavation for any footings, with any reinforcing mesh in place
- after a major concrete pour
- drainage lines, before connection (where waste water and septic tanks are used)
- building frame
- completion.

Progress payments can be scheduled in conjunction with these stages and with the delivery of material to the site.

Inspection of electrical connection is conducted separately. A final inspection is necessary before permission can be gained to occupy the building.

CHAPTER 2

Construction materials

Steel, concrete and timber are the common materials used in the construction of farm buildings. An introduction to their main characteristics will enable good decisions to be made about their selection, fabrication and installation for various construction projects.

Steel

Different types and grades of steel are in common use for farm buildings. The following terms describe their main physical properties:

1. Strength; refers to the ability to resist a static load, in tension and compression, before yielding or breaking.
2. Hardness; refers to resistance to abrasion or indentation.
3. Toughness; refers to a metal's resistance to shock loading, and repeated shock loading (fatigue).
4. Ductility; a related property of a metal which enables it to accept mechanical deformation without cracking. (It is the opposite of brittleness.)

All of these properties are influenced and manipulated by alloying, heat treatment (even inadvertent heat treatment such as welding), and mechanical working such as cold rolling. Other properties which may be important in certain applications are conductivity (of heat or electricity), weight and corrosion resistance.

Plain carbon steel

Steel is produced by remelting pig iron, or from molten iron direct from the blast furnace. Oxygen is introduced, which by a complicated series of chemical reactions, allows remaining impurities in the iron to be removed as slag. The carbon content is controlled precisely to give the required type of steel, by the introduction of the mineral anthracite into the melt. Normal commercial plain carbon steels still contain a very low proportion of other elements such as manganese, silicon, phosphorus and sulphur, normally regarded as impurities.

It is the carbon content of the steel, and the form or structure in which the carbon occurs, which determines the properties of the steel. Generally speaking, a higher carbon content (up to about 1.0 per cent carbon) increases the hardness and strength of the steel but decreases the ductility. Higher carbon content increases hardness further, but diminishes the strength. (When the carbon content reaches 2.5 per cent the metal is classed as cast iron).

Mild steel has a carbon content of around 0.2 per cent, and is widely used for steel plate, pipes and structural sections. Increasing the carbon content increases the strength of the steel and its response to heat treatment.

Medium carbon steels (0.25–0.45 per cent) are used for high strength, cold rolled sections, axles and shafts, hydraulic lines, hoes, discs and so on, whereas high carbon steels (0.45–1.0 per cent) are used where high strength and hardness are required; railway tracks, springs, anvils, self-tapping screws etc. An even higher carbon content produces steel used in knives, drills, chisels, axes and cutting tools.

Deliberate addition of other substances such as nickel, chromium, vanadium and molybdenum produces alloy steels with special characteristics. For example, the addition of chromium is the basis of stainless steel, where corrosion of the steel is prevented by a naturally occurring microscopically thin chromium-rich oxide film.

Strength of steel

If a piece of steel is subject to a tensile force (that is a force trying to stretch the metal), a number of things happen to the steel as the force gets greater. The most obvious of these is that the piece of steel will stretch, and then break, as the force applied exceeds the strength of the metal.

For commercial grade mild steel, the force that the steel can withstand before it breaks is around 405 MPa (megapascals), which is equivalent to 2600 tons per square inch. It can be much higher for high carbon steels, especially those with appropriate heat treatment. This load corresponds with what is known as the ultimate tensile strength of the steel, and it is easy to see why steel is such a popular construction material.

In any structure or machine, we do not want components to break (except for cases like shear pins on farm machinery, which protect the machine from overload). Therefore the ultimate strength of the steel must be greater than the loads which the components are asked to carry. When small loads are applied to a steel component, it deflects or stretches an amount

Figure 2.1 *Steel is a popular construction material. In addition to good strength-weight characteristics, its properties and dimensions are uniform and predictable. It is free of defects, and usually easy to work in most farm workshops. It gives a clean and straight appearance. The main disadvantages are associated with corrosion, heat transfer and noise*

corresponding to the size of the load, and returns to its original shape or position when the load is removed. In more technical terms, the applied load results in a certain amount of "stress" within the component, the corresponding deflection or stretching the component undergoes is called the "strain".

As the load increases, a stage is reached where the steel will not return to its original shape or position when the load is removed. The component is deformed permanently, and the steel is said to have yielded. For most purposes the component has failed, even though it has not broken. The stress or load at which this occurs is known as the "yield stress", and is typically around half the ultimate strength of steel; around 250 MPa (1600 tons per square inch) for commercial grade mild steel. Semi-high tensile steels, with higher yield strengths, are increasingly popular, since their greater strength enables lighter sections to be used.

Shaping

Many metal components such as structural steel components, pipe and tube, wire, rod bar and sheet, are produced by shaping solid metal, hot or cold. There are four principal methods of shaping:

1. Forging is a process of heating the metal until quite soft, then hammering it into the required shape, followed by appropriate heat treatment. This is the method employed by blacksmiths.

2. Extrusion is a process where heated solid metal is squeezed under very high pressure through a die, with the resultant product taking the cross-section of the die, in continuous lengths.
3. Hot rolling is used to break down large ingots into more useable shapes (plate, rod, bar), which can then be further fabricated into the final product. Rolling is used to produce structural shapes such as universal beams.
4. Cold rolling is rolling the metal whilst cold. It requires more energy to perform since the metal is harder, but allows more accuracy in fabrication, and provides a better surface finish. It also increases the strength of steel by physically altering and distorting the crystalline structure of the steel. This allows slightly lighter weight components to be used. It does however slightly increase brittleness. Steel cladding and many light-weight steel building components are cold rolled.

Corrosion

Corrosion is the destructive attack on a metal by, or as a result of contact with, agents such as rain, dung, urine, water carrying dissolved salts, polluted air, or aggressive chemicals. Rusting of iron and steel is the most obvious example of corrosion, but most other metals corrode, some to a lesser degree than others, and sometimes only under certain circumstances. Corrosion protection must be considered for all metal structures.The degree of protection is dictated by environmental conditions.

Various circumstances surround the process of corrosion. Generally speaking, corrosion is accompanied by the establishment of very small and localised electric currents which result in a chemical reaction between the metal and air or some other chemical substance. Corrosion is greatly accelerated when moisture is present because water is a good electrical conductor. Salty water is even more corrosive. These electric currents can become established by having two different metals in contact, by impurities in the metal, at cracks and other points of concentrated stress, or even by stray voltages caused by improper earthing of the electrical wiring within a building.

Some metals, for example aluminium, are corrosion resistant. Aluminium resists corrosion by forming an oxide layer on exposed components. The oxide layer is itself a product of corrosion but it protects the aluminium underneath from further corrosion. However, steel is liable to continue rusting because the rust (iron oxide) is porous, and it tends to flake off, allowing continuing exposure to corrosive forces.

The most common method to protect steel is to paint it; to coat the metal with an impervious layer. Provided the bare metal is prepared adequately, modern metal paints provide excellent protection with reasonable periods before repainting. Special primers and paints are available for situations where corrosion risk is higher than normal. Some products, referred to as "cold galvanising" paints, contain zinc.

Galvanising is accomplished by coating the steel with a thin layer of zinc, or zinc and aluminium, and is another very popular method for protecting steel. The zinc coating provides a different type of protection than does paint. An electric current is still present between the steel and the zinc, but the zinc corrodes in preference to the steel in a process known as "cathodic protection". The zinc will eventually corrode away, but should last many years depending on the thickness of the zinc layer and the conditions of service. If the galvanising becomes damaged and exposes the steel, the steel will corrode at that point, but this will probably not be as bad as corrosion occurring and spreading under paint.

Galvanising has a higher initial cost than painting, but provides a longer period of protection. It is essential where painting cannot be undertaken. Most steel structural components can now be purchased in galvanised form. Note that where these components are drilled, cut or welded, paint must be applied since the galvanising is broken at these points, and localised weakness may otherwise occur.

Nuts, bolts and other fasteners can be purchased galvanised, or plated with corrosion resistant alloys.

Structural shapes and sections

Structural steel is available in a range of standard shapes and cross-sections; pipe, rectangular hollow section (RHS), universal beams and columns, angle, channel, purlin material etc. A range of standard wall thicknesses is available. Check the lengths supplied. Most are available galvanised, either standard or at extra cost.

Refer to your local steel supplier for availability and cost. They may also be able to supply some information on the load or span capability of some items, but do not take this as specific engineering advice. They may also have some pre-fabricated structural items such as lattice trusses.

Concrete

Concrete is a mixture of cement, coarse aggregrate (such as blue metal), fine aggregate (sand, less than 5mm) and water. It is plastic when first made (that is, it flows quite readily), then hardens to a strong, homogenous mass possessing great durability.

The main reasons why concrete is such a popular and versatile construction material are:

- While plastic it is easily moulded into almost any shape, and a number of surface finishes.
- Once set it is capable of withstanding very large loads in compression. Its strength in tension is provided by inclusion of reinforcing material.

- It is extremely durable; it has excellent resistance to weathering, insect attack and fire.

Each of these properties is dependent firstly on the proportions and quality of the components used to make the concrete, and secondly by the way in which the mixed concrete is handled before it hardens.

Cement

Cement has a complex chemical composition. It is produced by the burning of limestone and special clays, which are then crushed into a very fine powder. The proportion of the component minerals can be varied during manufacture to give the cement varying physical properties.

When water is added to cement, a chemical reaction occurs which, after some time and provided sufficient water is present, results in the hardening of the cement paste. This chemical reaction is known as "hydration". This is not drying; in fact the concrete must be kept moist during the hardening process for the hydration reaction to proceed adequately, and to ensure maximum strength in the concrete. Compressive strength of concrete is measured in megapascals (MPa). Standard pre-mixed concrete has a compressive strength of 20 MPa (about 125 tonnes per square inch). Higher strength concrete is available for special applications.

Additives are sometimes introduced to provide for quicker or slower setting, to add colour, to provide a particular surface finish, or to improve the flow characteristics of the concrete.

Properties of plastic concrete

The properties of plastic, or wet, concrete that affect its use in the construction of farm buildings can be classed under workability and cohesiveness.

Workability

The workability of concrete is a measure of the ease with which it may be placed and compacted. Workability is affected by:

1. Water content: the higher the water content for a given amount of sand and aggregate, the more workable the concrete. However, an increase in water content without a corresponding increase in cement will increase the water : cement ratio, which is fixed by considerations other than workability (strength and durability).
2. Cement content: the cement paste in concrete acts as a lubricant. At a fixed water : cement ratio, the higher the cement content (and therefore the higher the water content) the more workable the concrete. Therefore, adjustments to the workability of concrete should be made by adding paste of the required water : cement ratio, and not by adding water alone to the sand or aggregate.

3. Aggregate grading: the grading of the aggregate also affects workability. Generally, those having a range of particle size ("well graded") produce more workable concrete.
4. Particle size and shape: rounded aggregates tend to produce more workable concrete than rough angular mixes.
5. Plasticising additives: these can be added to improve flow characteristics.

Cohesiveness
The cohesiveness of concrete is a measure of its ability to resist segregation and bleeding during placing and compaction. Segregation is the separation of the components. Bleeding is the migration of water to the surface.

Factors affecting cohesiveness are:
1. Consistency: very wet mixes slump readily and segregation of the cement paste from the coarse and fine aggregate occurs more easily. Very dry mixes are friable (easily crumbled) and again segregation may occur. Great care must be taken when handling very dry or very wet mixes.
2. Aggregate grading: mixes which are deficient in cement or fine sand aggregate tend to segregate readily during handling. The presence of an excessive amount of fine material tends to make the concrete sticky (very cohesive) and it is more difficult to work or place.

A formal measure of the suitability of the plastic concrete is the "slump test". A steel cone of specific dimensions is filled with concrete. The method of filling and tamping is also standardised. The cone is then turned upside down, and lifted off, allowing the concrete to slump. How much the free-standing concrete slumps is then measured; around 80 millimetres is about right.

Properties of concrete in the hardened state

Hardened concrete is very strong in compression, although comparatively weak in tension. The lack of strength in tension is overcome by reinforcing the concrete, and this will be discussed later.

The compressive strength of hardened concrete is dependent on a variety of factors as follows.

Aggregate
The strength of concrete is influenced by the size, and variation in size (the grading) of the aggregate components.

The water cement ratio of the mix
Excessive water in the mix reduces the strength of concrete. So does insufficient water, by not providing sufficient moisture for the hydration reaction. During a concreting job, it is sometimes tempting to add a little water as the work progresses, to maintain workability. However, this practice should be avoided, as it will affect strength. If necessary, add water and cement paste.

Better still, do the job in smaller, easier to manage sections, with correctly formed structural joints between adjacent sections.

Density

Good concrete generally has a density between 2.2 and 2.6 tonnes per cubic metre, depending on the density of the ingredients.

The density of concrete will be reduced by insufficient compaction. Voids (air pockets or pores) in the concrete created by excess mixing water or improper compaction reduce the density and the strength of the concrete. Voids of 10 per cent will reduce strength by 50 per cent. Voids are often associated with placement of concrete around awkward reinforcing layouts, or unusually deep formwork.

Tamping of the concrete, with a rod or special vibrator, is necessary in such situations, but there is a risk of segregation if excessive vibration is used.

Age

Properly cured concrete will continue to gain strength for many years, although most of its eventual strength is gained in the first month.

Concrete gains strength rapidly at first and more slowly as it becomes older. It can usually be walked on after 24–48 hours. On some jobs, specifications require no work on the concrete until after a particular time period, often seven days. This would apply mainly to above ground concrete construction.

Curing conditions

To gain its ultimate potential strength, concrete must be kept moist during curing so that water is available at all times to react chemically with the cement.

The moisture which accumulates on the top of the concrete shortly after screeding will evaporate fairly quickly. Do not sprinkle cement powder on it to absorb it. Cover the slab with plastic sheet to prevent moisture escaping, or sprinkle water on it at frequent intervals. Spray-on curing compounds are available.

Temperature will also affect the rate of curing. Try and avoid hot times of the day. Working in excessive heat also results in a rushed job, since finishing conditions are reached and passed too quickly. Also avoid very cold weather.

Pre-mixed concrete

Pre-mixed concrete may be completely mixed in a central plant mixer and transported to the site in a truck agitator/mixer, or it may be truck mixed.

With truck mixed concrete, the separate materials are usually mixed dry at the concrete batching plant. Water may be added at the plant, during

Table 2.1 *Home mix concrete proportions*

Type of work	Nominal mix proportions (by weight) cement: sand: coarse aggregate	Materials required per m³ of concrete cement: sand: coarse aggregate (bags, m³, m³) (measured loose)	Added water per bag cement (L) (maximum)
Structural concrete and wearing surfaces	1 : 2 : 3	8 : 0.65 : 0.7	21
Unreinforced concrete (mass footings, light duty paths, driveways etc).	1 : 2.5 : 4	7 : 0.7 : 0.7	24

Note that the moisture contained in bulk sand will influence its weight, and the amount of water to be added.

transportation from water tanks on the truck, or at the site. This is the most common type of mixing. The inclined drum mixers as used on trucks have the mixing blades so arranged that when the rotation of the drum is reversed, the concrete is discharged.

Any special requirements should be discussed with the supplier. These might include the need for a high strength mix, unusual delivery arrangements, or the need for slow setting concrete.

Home made concrete

Although most concrete used today is ready mixed and ordered for the required strength and properties, we all mix a little ourselves from time to time for minor works. Table 2.1 gives a guide to mix proportions for a 20 MPa strength concrete using 20 mm aggregate. For low duty concrete, a 15 MPa strength may be satisfactory.

Typical concrete practice

Consider the construction of a concrete slab for a shed which is to be built on a near level site. Site preparation involves levelling the area under the slab, taking due precaution for drainage around the shed, and removing the topsoil to ensure a stable foundation. On larger jobs it is necessary to excavate some distance so that concrete footings can be constructed on stable foundation material.

The thickness of the slab is determined depending on the load to be carried by the shed floor (100–125 millimetres is typical). Formwork is then erected. In this example the formwork would consist of timber boards, straight in both directions, which are vertically located around the exact perimeter of the slab site to form the edges of the mould. Steel or wooden pegs are used to hold the formwork in position. In addition the top edge of the formwork is located at exactly the height and level of the proposed slab, to facilitate finishing. The formwork must be held securely to prevent it moving when the concrete is poured.

One minor disadvantage of concrete is that it is slightly porous. A plastic membrane is often required underneath the slab, to prevent moisture rising upwards through it. A layer of washed sand or fine aggregate is placed under the membrane to protect it from being punctured by small stones, and to help provide a level surface onto which the concrete can be poured. This is particularly important in buildings, but not in all applications.

Steel reinforcement is then positioned. Although plain concrete is very strong in compression, it is relatively weak in tension, and so steel reinforcing rods or mesh are located within the slab at positions which correspond to the location of tensile forces in the structure, and to control shrinkage. For our building slab, this is near the top, but there must still be at least 25 millimetres of concrete above the reinforcing, mainly to protect it from

Figure 2.2 *Concrete being poured from the truck discharge chute. Note the formwork to define the perimeters and level of the slab and the steel reinforcing in place*

Figure 2.3 *Screeding wet concrete to a rough level*

corrosion. The reinforcing mesh is positioned with wire supports, themselves located on small metal trays to prevent puncturing the plastic membrane. Plastic "chairs" can also be used. If holding down, or anchoring bolts (bolts protruding from the slab, to which the frame of the shed is attached) are to be used, they should also now be placed. Ensure they are accurately positioned and secured. They may even be attached to the reinforcing.

The concrete is then poured into position, using shovels to spread it within the formwork. It is then "screeded" to provide a level, but somewhat rough surface. Screeding involves moving a metal straight edged bar across the concrete, with the ends of the bar resting on the top of the formwork, and pushing or pulling concrete in front of it. The concrete surface will end up on the level set by the top of the formwork.

The final finish to the surface is performed by floating or trowelling, when the concrete can just take some weight.

Handling concrete

Concrete must be handled in a manner to prevent segregation which can occur when unsuitable methods are used to transport, place or compact the plastic concrete. Segregation results in the hardened concrete being non-uniform, with weak and porous honeycombed patches.

Placing plastic concrete
- Concrete should be placed vertically, as near as possible to its final position. Spreading, if necessary, would be done by shovel and not by causing the concrete to "flow" with mechanical vibrators.
- Concrete should not be dropped from an excessive height. If the height will be more than one metre, use a chute.
- Start placing concrete from the corners and from the lowest level.
- Each load should be placed into the face of the previously deposited concrete.
- Where a layer of concrete cannot be placed before the previous layer hardens, as for example on the morning after an overnight stop, a construction joint should be formed.
- Concrete should not be placed in heavy rain unless there is overhead shelter, otherwise the rain may wash cement from the surface.
- Access to the site is important, to avoid carting the concrete from the delivery truck to the site in wheelbarrows, and to keep the time the truck unloads to about 20 minutes.

Compacting plastic concrete
The object of compacting concrete is to ensure that maximum density is obtained and that complete contact with the reinforcement and formwork is achieved.

Thorough compaction is most important as it leads to:
- maximum strength
- sound, more watertight concrete
- sharp details at corners
- good surface appearance
- good bond with reinforcement
- adequate protective cover to reinforcement.

CONSTRUCTION JOINTS

Construction joints are used when the pour is interrupted for some reason. Temporary formwork is constructed and used to enable proper compaction of the first part of the slab. Where reinforcing steel is being used the steel should protrude through the join to allow bonding into the second part of the pour. This means the temporary formwork must be constructed around the exposed reinforcing.

Before pouring the second part of the slab, the aggregate of the first part should be exposed to help achieve adhesion across the joint. This can be done when it has sufficiently hardened.

Note that a construction joint will always be a point of weakness in the slab, where cracking is possible. They should be located where the effects of this will be minimised. Sometimes the joint must be specially designed to accommodate expansion and contraction.

Ordinary hand methods of compaction consist of rodding, tamping and spading with suitable tools and are satisfactory for small jobs. Mechanical vibrators are required where large volumes of concrete are to be used.

The most often used concrete vibrators (and probably the most effective) are called "immersion vibrators" or "poker vibrators". They consist of an eccentric shaft which rotates in the head of the vibrator, driven by a flexible shaft from a small portable petrol engine.

It is most important not to over-vibrate the concrete as this causes segregation. The vibrator should always be inserted into the concrete vertically and slowly withdrawn when mortar appears on the surface, and bubbles stop rising.

Finishing concrete

The satisfactory finishing of concrete requires that all of the steps leading up to the finishing — mixing in the correct proportions, placing and compaction — be properly executed.

Irrespective of the type of surface finish required the essential requirements are that initial finishing should be completed as soon as possible after placing and vibration. Final finishing — floating and trowelling — should be delayed until the surface is ready.

Typical finishing procedure for a slab is to screed the concrete to a level but rough surface while the concrete is still plastic. Bleed water will come to the surface as the concrete "goes off". No finishing operations should be performed in any area where there is free surface water.

When the sheen has left the surface the floating operation takes place. Floating is the operation of smoothing irregularities in the surface following screeding. Its purpose is to:

- embed large aggregate below the surface
- remove imperfections
- provide a denser and smoother surface
- close minor surface cracks which sometimes occur as the concrete sets hard.

A wooden float produces a rough textured surface which is non-skid and safer for paths and driveways than that of a trowelled finish. A broom finish is sometimes used.

A steel trowel is used to provide a smooth, dense and hard surface. This type of surface is durable and easy to clean, and hence is used on factory and workshop floors. It can be slippery when wet. Power trowels are in common use.

The first trowelling may produce a sufficiently fine surface, but additional trowelling may be used to increase smoothness and hardness if desired. Pressure is exerted on the trowel to compact the paste and form a dense, hard and durable surface.

Figure 2.4 *Power trowels make quick work of finishing larger concrete jobs*

Curing the concrete

All of the above hard work can easily be wasted if the finished product is not adequately cured.

The properties of concrete improve with age, provided conditions are favourable for continued hydration of cement. The improvement is rapid at early ages but continues more slowly for an indefinite period. However, two conditions are required:

- the presence of moisture
- a favourable temperature.

Evaporation of water from newly placed concrete can cause the hydration process to stop. Loss of water also causes concrete to shrink, creating tensile stresses and hence cracking of the surface.

Hydration proceeds at a much slower rate when temperatures are low. Practically no hydration takes place when the temperature is near or below freezing.

Concrete should be protected so that moisture is not lost during the early hardening period and concreting jobs should be avoided during very hot and very cold weather.

Reinforced concrete

Good quality concrete is very strong in compression, but relatively weak in tension. To overcome this problem reinforcement is placed in the concrete, located in areas of tension.

With slabs on the ground, which are theoretically fully supported, the steel reinforcement is placed nearer the surface of the slab to control shrinkage cracks. In a suspended slab or beam between two walls the reinforcement is placed near the bottom, or the tensile face of the slab. Note that the reinforcing may not eliminate cracking, but rather control its extent.

Location of reinforcement is important. It should be supported on "chairs" to ensure correct location. The common practice of placing the reinforcement part way through a pour should be avoided. Make sure there is good contact between steel and concrete. Excessive rust should be removed from the steel.

For concrete that is fully supported, and reinforcement is needed to control shrinkage, polypropylene fibres can be used for reinforcing rather than steel, included in the concrete during mixing.

Timber

Although steel is probably preferred for structural purposes and economy, timber is in widespread use in older buildings, and where its unique properties make it suitable for particular applications.

Hard and soft woods

The main visual distinction between hardwood and softwood is that hardwoods have "vessels" and softwoods do not. Vessels are much larger in cross-section than the other wood elements and can be readily seen with the unaided eye. Vessels are also called "pores" and the terms "pored" and "non-pored" woods are often used for describing hardwoods and softwoods respectively.

The terms hardwood and softwood are misleading as they bear little relationship to the relative hardness of the wood. Balsa, for example, is botanically a hardwood. Australian hardwoods are generally classified as a group, perhaps distinguished by their region of origin.

All the eucalypts, acacias, oaks and maples are examples of hardwoods. Douglas fir (Oregon), cypress pine and radiata pine are classified as softwoods.

Sapwood and heartwood

The wood of the tree is differentiated into two zones, the outer "sapwood" and inner "heartwood". Sapwood is generally lighter in colour, less durable and when freshly cut from a tree, has a higher moisture content. Heartwood has no living cells whereas the sapwood has some living cells. The change from sapwood to heartwood takes place when the living cells die.

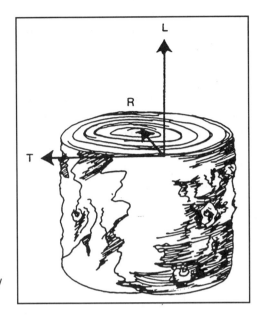

Figure 2.5 *The properties of timber vary with the grain.*

Sapwood is as strong as heartwood but heartwood is more durable since it is less likely to be attacked by fungi and insects. Most trees have a comparatively narrow sapwood, 10–50 millimetres in width.

Sapwood is, however, more readily penetrated by preservative fluids than heartwood and so properly treated can be rendered quite durable.

Cell arrangement and strength

Wood is not "isotropic", it does not have the same properties in every direction. This is because the cells are oriented in the direction of the tree height. It differs in strength and other properties in the longitudinal (L), radial (R) and tangential (T) directions (see Figure 2.5).

For most structural purposes, it is necessary only to differentiate between directions parallel or with the grain (longitudinal) and perpendicular to, or across, the grain (radial and tangential). The exception is shrinkage, which can be significantly different in the radial and tangential directions. As a rough guide, the relationship of tangential to radial to longitudinal shrinkage is about 100 : 50 : 1.

Factors affecting strength

The principal factors affecting operational strength are timber species, variation within species, defects, moisture content, temperature and duration of load.

Species

The strength of different species of timber varies enormously, some being several times stronger than others. This strength difference is further accentuated by defects in the timber, some species being more prone to defects than others.

Variability within a species

Even within a particular species, some pieces of timber may be three or four times stronger than others. The variation in strength within a species, other than those caused by defects, can result from one or more of the following:

1. Density. Generally the denser the timber, the stronger. Younger trees tend to be less dense than older trees.
2. Rate of growth. Timber of medium growth rate seems superior to that of very high or very low growth rate.
3. Position in the tree. Generally the wood near the butt of softwood trees is stronger than that near the top. In hardwoods (including most eucalypts) the opposite is the case.
4. Condition of growth. Height above sea level, rainfall, temperature, soil type, spacing of trees and various other environmental factors all have an effect on the density and therefore the strength of the timber, although there is little scientific evidence on their specific influence.

Defects

A serious defect can reduce the strength of a piece of timber to practically zero, irrespective of species, moisture content and so on. These severe defects can usually be detected visually and the timber cut accordingly or discarded.

Serious defects affecting the strength of timber are as follows:

1. Knots. They lower the strength of timber, particularly in bending. Knotting is common in softwoods, such as Douglas fir (Oregon), radiata pine and cypress pine, but are less common in Australian hardwoods supplied for structural purposes.
2. Sloping grain. Slope of up to 1 in 20 should not appreciably alter the strength, but a slope of 1 in 8 can lower the strength by half.
3. Decay. Badly decayed timber has very little strength and should not be used structurally. Heart rot in green durable eucalypts rarely continues in the sawn timber.
4. Insect attack. Damage varies from negligible to complete destruction depending on the extent of attack. Termites (white ants) are a serious menace in some districts but attack is generally prevented by suitable construction techniques, preservative treatment of the timber or poisoning of surrounding soil.
5. Other defects include compression failure, checks, shakes, bark or sap inclusions, splits and warp (see Figure 2.6).

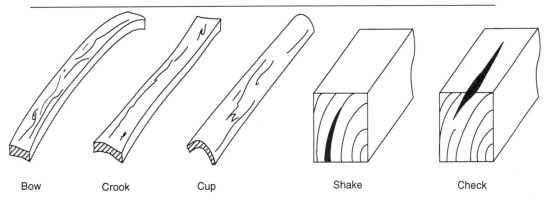

Bow Crook Cup Shake Check

Figure 2.6 *Typical defects in timber*

Moisture content

The strength of wood is greatly influenced by moisture content. Green timber may have a moisture content as high as 100 per cent, doubling the dry weight of the timber. However there is very little change in strength characteristics until the moisture content drops to below about 30 per cent: the fibre saturation point. Below this point, strength increases markedly, so that at 12 per cent moisture, the strength may be 75–100 per cent greater than at 30 per cent moisture. Generally seasoned timber will have a moisture content of between 10 and 15 per cent.

In larger cuts of timber, the increase in strength due to drying is partially offset by the weakening effect of defects from drying, such as checks. An increase in strength of 25 per cent over green timber is usually allowable.

Figure 2.7 *The vertical boards around this round yard were butted together and fixed when green. Shrinkage of the timber has caused significant gaps between the boards*

Shrinkage occurs during seasoning, so its always best to season structural timber under controlled conditions before use.

It is important that in all structures made from green timber the possible effect of shrinkage be taken into account, particularly at the joints. For example, when using green cypress pine tongue-and-groove flooring, the boards are cramped up very tightly together before being nailed to the floor joists. This is because the boards shrink as they dry and the joints open up. The cramping reduces the eventual gaps between the boards. On the other hand, if kiln-dried radiata flooring boards are used, the boards should only be "touch tight" on nailing since they will take up moisture from the atmosphere and swell. If they are cramped this swelling would cause the floor to bow up off its supports.

Ideally the flooring should be left for a reasonable period on the joists before fixing so as to attain their working moisture content for the area.

Temperature
The strength of timber decreases with increase in temperature. Permanent damage does not occur under normal climatic variation, but may occur if wood is held at high temperature for long periods, especially if the initial moisture content is high.

Duration of load
The mechanical properties of wood are considerably affected by the duration of loading; the shorter the period of loading, the greater the load that can be carrried. Repeated application of a temporary load can have a cumulative effect, and may have the same effect on the structure as if the member were continually loaded.

Duration of loading affects the stiffness as well as strength. A green timber beam tends to sag under load so that the deflection of the beam after one year may be several times the initial deflection. This is an important consideration when calculating the required size of timber beams.

All of the above factors contribute to a lot of variability in strength of timber, making it a little less predictable for structural purposes compared to steel.

Selection of timber for structural purposes

There are two aspects needing careful consideration. Firstly the timber species that is to be used, including its quality and treatment prior to use, and secondly the size timber that will be needed.

Information is readily available on the selection of timber species, although often the final choice is only between hardwood and softwood, unless some particular requirements need to be met.

Each timber is allocated to a Strength Group, from S1 to S7 in the green condition, and SD1 to SD8 in the dry condition. Within each Strength Group, it is further divided into a particular Stress Grade, F4 to F34, which takes into account defects in the timber and their influence on strength.

The Stress Grade to which a particular timber belongs corresponds directly with its strength in bending, and accordingly, the maximum load that can be applied to the timber when it is used in construction. The higher the F-number, the greater the strength. There are two methods for stress grading. Firstly by mechanical means and measurement of strength of a sample of the timber, and secondly and more commonly by visual means. For example, "select grade" timber is free from defects and consequently of greatest strength for that particular timber, whereas in "standard grade" timber there are limited defects so that the weakest piece shall have at least 60 per cent of the strength of the "select" timber.

It is worth noting that it is not economically justified to use defect free timber for most structural purposes. Since only a small proportion of timber is entirely clear, it is quite expensive, and used mostly for special purposes. It is more practical to accept defects, but to place definite limits on the size and number that will be permitted. Laminated timber beams overcome this limitation by confining the defect to one lamination, not across the major part of the cross-section. Consequently laminated beams can be used in long span or high load applications where standard beams would need to be so large as to be impractical.

Figure 2.8 *Widespread use of timber in this shearing shed - timber beams, columns and purlins; tongue and groove flooring; plywood and particle board wall lining. Despite some possible limitations in regard to strength and durability, timber is preferred by some people because of its appearance and suitability to the application. It can also assist with insulation and dampening noise*

Information is also available regarding other factors such as density, durability, percentage shrinkage, susceptibility to borers, availability and typical applications. (Density refers to the weight of the timber per unit volume. Durability refers to the natural resistance the timber has against deterioration from fungal and termite attack.)

Preservation of timber

The main contributors to timber deterioration are decay, termites, borers, weathering and fire.

Decay

Decay or rot is the result of attack by fungi which penetrate the wood and use some of the substances in it as food. Because the fungi also require moisture, decay usually occurs in timber which is at a moisture content greater than 20 per cent, and so the problem is usually confined to timber exposed to the weather, posts and poles in contact with soil, or timber located where moisture collects through condensation or poor drainage.

To minimise decay:

- exposed timber should be protected with preservative or paint
- damp-proofing must be used in building construction
- airspace and ventilation must be provided under timber floors
- timber awaiting use must be stacked correctly
- correct drainage must ensure moisture does not remain in contact with timber.

Termites

There are many different species of termites, not all of which attack timber used in buildings, but those that do can create substantial damage. Termites are subterranean; that is they remain buried beneath the soil surface or enclosed within the wood they are consuming. When termites need to travel above ground they construct mud tunnels or covered walkways, which is often the first sign noticed of termite attack. They have the ability to travel into the timber of a building through cracks in concrete, under concrete slabs and through mortar between bricks, usually through those parts up to half a metre below the ground surface. Posts and poles in the ground are particularly susceptible.

Timber and the soil under and around buildings can be treated with chemical preservatives — standard procedures wherever there is a risk of termite attack. Ant capping does not stop termites, but enables their tunnels to be seen. Do not disturb the tunnels — obtain expert assistance for treatment.

Borers

There are also many different types of borer, and they mostly attack the sapwood of timber posts and poles. The starch content of the sapwood also partly determines the borers' appetite for it, and susceptibility to borer attack varies with timber species. Preservatives are used to render the sapwood resistant to attack.

Weathering
Alternate wetting and drying of exposed timber causes a change in size and cracking of the timber as a result of repeated swelling and shrinkage. Oil based coatings are suitable to prevent the migration of water into the timber, but because they also prevent the migration of moisture out of the timber, they can only be applied when the timber has been properly seasoned.

Fire
Heavy timbers are often difficult to ignite, but this does not mean that the hazard should not be considered. Special protection of timber itself is usually not warranted, and fire prevention and control is usually provided for farm buildings by other means.

Wood preservatives

There are a number of preservatives available for timber treatment, classified as either oil based or water based, and the most common of each are described below.

Creosote is an oil derived from the distillation of coal tar, which is quite effective and widely used. Sump oil has been used as an alternative to creosote, but it is not as effective and it reduces the effectiveness of creosote when used to dilute it. Care should be used when handling creosote because it can irritate the skin. Workers should have personal protective clothing when using it.

Creosote can be applied in the field by a cold soaking process, which is better than painting. This involves first seasoning the timber, which requires at least three months of dry weather to reduce the moisture content of the sapwood, then placing the timber upright in a tank of creosote (a 200 litre drum will do) for about six days (longer in cooler weather). An alternative method for water soluble preservatives is to stand the timber in a drum of preservative immediately after removing the bark, and allowing the preservative to replace the sap as it evaporates from the exposed surface of the timber.

CCA treatment involves use of a preservative consisting of a mixture of copper, chrome and arsenic salts. These salts are soluble in water before and during treatment, but are converted to insoluble compounds in the timber. The green colour of timber treated with this chemical is a distinguishing feature. The chemical is applied to the timber with specialised equipment which forces the preservative into the timber under pressure.

Fasteners

Fasteners deserve special mention. There are hundreds of different types of fasteners used to attach wood to wood, wood to metal, metal to metal, and both wood and metal to concrete. It is impossible to describe them all, and their characteristics and applications, so this section only presents a sample of the different types available.

Structural fasteners

Fasteners often perform the most critical function in a structure; that of holding it all together. Yet the use of most types of fasteners involves drilling holes or otherwise interfering with load bearing components, and so tends to produce a point of weakness in the structure where loads are greatest. It is often found that failure of a structure occurs at the points of connection within it. Fasteners are important for the potential weakness they cause as well as their potential strength.

In timber construction, bolts are commonly used to join structural components. The diameter and number of bolts is determined according to the maximum expected load on the join, and should be located so as to spread the load over as much of the timber cross-section as possible. Figure 2.9 indicates that there is also a requirement to keep the bolt holes a certain distance from the edges of timber components, because of the risk of the timber splitting. Bolted connections should be tight, to provide some timber to timber friction to assist in withstanding the load. Washers of a suitably large diameter must also be used to avoid crushing the timber under

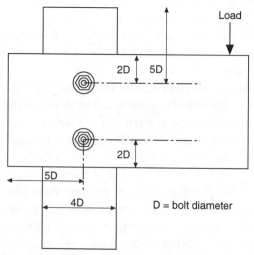

Figure 2.9 *Typical minimum requirements for bolted connection in timber. If the cross-section is large enough, the alignment of the bolts should be staggered*

Figure 2.10 *Pneumatic nail guns do a better and quicker job at repetitive timber nailing, such as for the floor battens for this shearing shed. Timber splitting is greatly reduced and a more secure join is achieved in a fraction of the time*

the bolt when it is tightened. It is preferable to use multiple bolts, spread over the cross-section of the timber, to spread the load.

Metal nailing plates are commonly used in timber roof trusses. The nailing plate, which is available in a multitude of sizes, consists of a galvanised steel plate with a large number of pointed slots punched into it, the metal protruding at right angles forming "nails". The sharp points on these protrusions enables the plate to be hammered or pressed over a join, providing great rigidity and strength easily and quickly. Home made gussets can replace commercially available nailing plates, but are slower to use and probably no less expensive.

There are a large range of specially designed timber fasteners. Framing anchors and joist hangers are mostly used in domestic house construction. Although less common, devices such as split ring connectors, bulldog connectors and shear plates could find a place in construction of farm buildings using heavier timber components. In bolted connections, there is a risk that the relatively narrow cross-section of the bolts may impose an excessive crushing force on the end fibres of the timber in the connection. These special connectors are designed to spread the load at the connection over as much of the timber area as possible, thereby creating a stronger join which is less liable to give and stretch under load.

With the popularity of steel construction for farm buildings, welding is a suitable means of joining components. However, a bad weld as a result of

Farm Buildings

TRIP-L-GRIP

Manufactured from 1.2mm galvanized steel, standard Trip-L-Grip connectors are amongst the most versatile of all timber anchors.

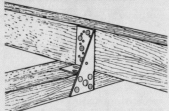

Typical applications include rafter to plate, joist to plate, roof truss to plate and joist to trimmer. When used correctly with the specially developed TECO nails, rigidity and resistance to short and long term loads is impressively high.

UNIVERSAL TRIP-L-GRIP

Another TECO development evolved from the standard Trip-L-Grip, the Universal is supplied unbent and can be formed on site to suit specific applications.

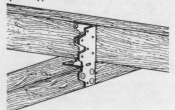

Like most nail type timber fasteners, this connector is designed to load its fastening nails in shear, substantially increasing resistance to withdrawal.

TYLOK

Tylok plates were originally designed for on site production of roof trusses as they can be applied by a hammer as well as an hydraulic press. Tylok plates are suitable for a wide range of applications in timber joinery such as trusses, formwork, site splicing, etc. They are also available as internal or external corner connectors where they are used for pergola construction, pallet repair and large timber case construction.

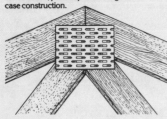

JOIST HANGERS

TECO Joist Hangers are designed where a strong rigid joint is required between two members meeting at right angles. Typical applications include joist to beam, beam to beam, rafter to beam.
Available in sizes from 40 x 90mm to 50 x 190mm, they can carry loads up to 3.5kN, depending on timber type.

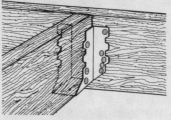

POST ANCHOR BASE

The TECO Post Anchor Base is designed to provide efficient and economical anchorage of wood posts to concrete slabs. Engineered to resist uplift resulting from high velocity winds.

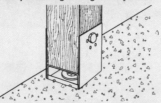

Bottom of post is kept above ground level so that there is no contact with dampness. Can be adjusted after installation. Suitable for fixing to existing or new concrete. Can resist uplift of 6.0kN (600 kg).

TRUSSBEAM

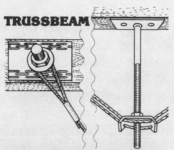

TECO Trussbeam is a unique twin wire system for support of purlins where it is difficult to use props or struts. They can also be used to increase the span of hips and rafters where large timber sizes are either too expensive or just not available. Trussbeams are manufactured in a wide variety of lengths from 2000mm to 8000mm.

WET POST BASE

TECO Wet Post Anchor Bases are designed to provide efficient and economical anchorage of nominal 100m x 100mm or 100mm x 150mm wood posts to concrete slabs or piers. Engineered to resist uplift resulting from high winds, they incorporate specially configured "anchoring arms" which are inserted in the concrete while still wet, eliminating the need for anchor bolts.

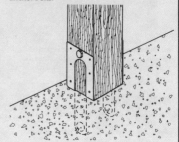

LOG DOGS

A variety of useful modular log structures can be simply and quickly assembled using the TECO Log Dog connector. The fastener is a rigidly formed galvanized metal cradle which will accept the natural variations present in log material ranging from 90mm to 140mm diameter. The connector is versatile in its applications and is quick and easy to install.

PUNCHED STRAPPING

Developed from galvanized hoop iron, TECO punched strapping is produced in thicknesses from 0.8mm to 1.2mm and in a variety of lengths. Punched strapping is a versatile general purpose building product particularly useful in small framing and roofing applications. It can also be used in a number of handyman applications around the home.

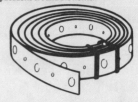

Figure 2.11 *Extract from the product information of one manufacturer of timber connectors.(Courtesy Gang-Nail Australia Pty Ltd.)*

Construction materials

JOIST STRAPNAIL

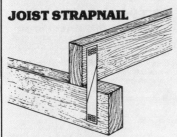

Used to connect beams which cross over one another at right angles (e.g. connection of hanging beams to ceiling joists) features pre-punched teeth for quick and simple application.

BATTEN TIE

Galvanized steel timber connectors utilizing integral teeth. Fast fixing of battens to rafters or battens to trusses to withstand cyclonic wind forces. The batten tie, a TECO development, is available to suit either 50 or 75mm material.

BULLDOGS

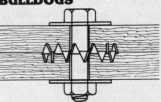

These multi tooth connectors are designed for use in soft timbers only. They are imbedded into the timber by tightening the central bolt. Joist strength is increased by shear transfer of teeth and bearing bolt enabling the use of smaller timbers and less bolts per joint. No joint preparation is required.

SPLIT RINGS

Available in two sizes, 64mm and 102mm nominal diameter, these rings are sprung into slightly larger grooves, cut into the adjoining faces of the members to be joined.

Each ring has an internal bevel, and is split to allow it to expand as the securing bolt through the centre is tightened, forcing the ring into the groove and locking it tight. Applications include the assembly of clear span timber trusses in large storage sheds, factories, bridges and timber towers. Advantages include the elimination of notches, keyways, steel tie rods and plates, thus reducing costs.

BRESSUMMERS

TECO Bressummers provide an effective roof load support over door and window openings and subject to selection of type can be used with either pitched roof construction or trusses.

TECO bressummers overcome shrinkage, warping and twisting and improve frame rigidity. Pre-punched, they can easily be fixed to the top plate and their strong rigid steel construction minimises deflection under load, reducing the likelihood of plaster cracks.

FOIL

The highly reflective foil faces of TECO Duralfoil — installed in conjunction with still air spaces of from 25mm to 100mm during construction of walls, ceilings and floors — will economically reduce heat loss or heat gain in a building. The ability of TECO Duralfoil reflective insulation to reflect light is a further benefit when it is left exposed as a sarking or underlay beneath the roof of factories, warehouses, etc. The effect is to make the best use of available light sources by reflecting light back into the work area, and give economies in the need for artificial light.

SPEEDBRACE

Speedbrace is a TECO patented tension bracing system that uses a pre-punched shallow 'Vee' shaped member that is easily handled and erected. Speedbrace can be used for both roof truss and wall frame bracing in low wind speed or cyclonic areas. It does not interfere with battens or purlins and cutting of structural members is completely eliminated. No tensioning devices are required.

BRICK VENEER TIES

These galvanized ties are available in a variety of shapes and sizes to suit all bricklaying requirements in brick veneer construction.

These include face of wall ties, which are fixed to the face of timber stud wall when insulation foil is used, and twisted veneer ties which are fixed to the side of timber studs when insulation is not required.

SHEAR PLATES

Made from malleable cast plate designed for wood-to-steel application. The circular plate is recessed into the timber member and a bolt with a large washer on each end runs through both members and the centre of the plate. This device increases the load capacity of the bolt by increasing the bearing area of the connector on the timber.
Shear plates are used in glue laminated beams and arches, fastening columns to footings through steel traps, and steel gusset plates to timber members.
Available in 68mm and 102mm nominal diameters.

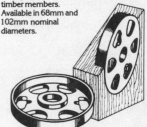

poor techniques, such as undercutting of the parent metal or inclusion of slag in the weld, will not be strong enough. Frequently the main frame components are bolted together, with brackets welded to them to accept the purlins and girts (these are the horizontal parts of the roof and walls respectively, to which the cladding is attached).

Fixing cladding

A wide range of fasteners is available for attaching the cladding to the building frame. Galvanised roofing nails are still quite acceptable when timber purlins and girts are used, but tend to become loose as the timber shrinks or during the movement of the frame. Nails with twisted or deformed shanks minimise this problem.

When attaching cladding to metal purlins and girts, self-drilling or self-tapping screws are used. They have also virtually superseded nails where timber is used. These screws are inserted with a special attachment fitted to an electric drill, so they are quick and easy to use. They have a sharpened drilling point and a threaded cutting shaft, so that they cut their own hole and thread as they go to provide a very firm connection. A wide range of types is available to suit most applications (see Figure 6.2).

Fasteners in concrete

A range of fasteners is available for fixing into concrete. A selection is shown in Figure 2.12.

Fasteners such as Dynabolts® are very popular for general purpose applications, and a wide range of special purpose fasteners is available. Fasteners of this type expand sideways against the wall of the hole, creating stresses in the concrete, so they need to be a certain minimum distance from the edge of the concrete. Chemical adhesive fasteners overcome this limitation, and are increasingly popular for slab edge fixing, installing reinforcing dowels etc. Powder actuated fasteners are also in common use, although a licenced operator is required.

Figure 2.12 *A selection of fasteners for fixing to concrete, showing installation method. (Courtesy Ramset Fasteners (Australia) Pty Ltd.)*

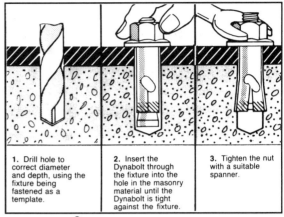

1. Drill hole to correct diameter and depth, using the fixture being fastened as a template.
2. Insert the Dynabolt through the fixture into the hole in the masonry material until the Dynabolt is tight against the fixture.
3. Tighten the nut with a suitable spanner.

DYNABOLT®

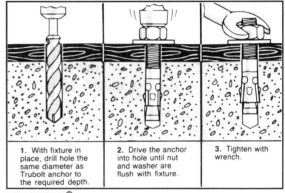

1. With fixture in place, drill hole the same diameter as Trubolt anchor to the required depth.
2. Drive the anchor into hole until nut and washer are flush with fixture.
3. Tighten with wrench.

TRUBOLT®

CHEMSET®

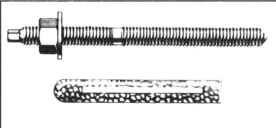

1. After the correct dimension hole has been drilled in the substrate
2. and thoroughly cleaned,
3. the capsule is inserted in the hole.
4. Using the appropriate adaptor fitted to a Rotary Hammer or Percussion Drill, the Stud Bolt is rotated into the capsule, splayed end first.
5. Under vibration/rotation of the Stud Bolt, the glass capsule is crushed and ground, mixing with glass, aggregate, resin and hardener, flowing around the Stud Bolt and against the sides of the hole.
6. The resin mix then hardens to chemically weld the Stud Bolt to the substrate.

IMPORTANT
The Stud Bolt must be inserted into the capsule under Rotation to ensure thorough mixing.

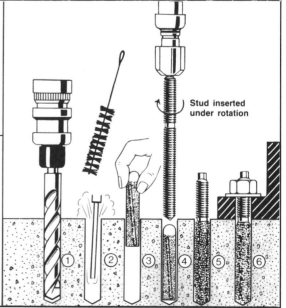

51

CHAPTER 3

Site measurements and plans

Building projects are designed and constructed from plans, which are presented in a standardised format to ensure all parties can understand the information presented. Plans represent the method of communication between designers, builders, owners and inspectors. It is important to understand the information contained in a plan, and how to read and interpret them. This will reduce the chance of problems during all stages of the project. It will also help ensure that the completed project looks and performs as you expect it to.

Accurate measurements are an essential part of plans and planning. Measurement requires consideration in two ways. Firstly, the site for the project needs to be measured, for horizontal distances and elevations. This enables the distances, areas and heights of the project site to be calculated. This is essential information during the planning and design stages. The preferred design is then drawn. Secondly, at the time of construction, all dimensions and elevations for the building are read from the plan, enabling the builder to position the building exactly as required and size all components correctly.

A good plan is also required to obtain the necessary approvals and to enable accurate quotations to be prepared. A confusing plan, or one containing insufficient information, will cause delays in the project, and possibly result in costly mistakes and adjustments during construction.

General measurements

Measurement of distances, elevation and angles will be discussed specifically later in this chapter, but some general comments apply to all of them.

Instrument selection

Various instruments are used to measure distances, angles and elevations in the field. The type of instrument used depends on a number of factors:
- the degree of accuracy required
- the number of measurements required
- the size of the area they are spread over
- the availability of the equipment.

For example, it is best to use a surveyor's level for measuring and setting out levels, but for small jobs that do not need high precision, there are other ways of doing it.

Drawings

There are generally two main types of drawings that make up the plans for a project. The first is a site plan or map which is used to show the general location of the building within the property, or part of the property. The scale of this drawing is typically between 1:500 and 1:2000, depending on the size of the project, (and the need to represent the site on a reasonable size sheet of paper). On a site plan, measurements can only be read to the nearest 0.1 metres, so it is allowable to use an instrument that reads only to that level of accuracy.

A building plan, however is drawn to a larger scale, typically 1:100. Measurements must be read to the nearest centimetre or millimetre (0.01 or 0.001 metres). This requires the use of instruments and techniques that provide the required accuracy.

A drawing showing the detail of some intricate part of the construction may need to be at a scale of 1:20 or larger.

It is important to be able to convert a measurement on the plan to a measurement on the site. If the scale of the drawing was 1:100, then each millimetre on the drawing would be 100 millimetres on the site. If the scale was 1:500, then each millimetre would be 500 millimetres on the site, and so on. When drawing a scaled plan, use a standard scale (such as 1:100, 1:200, 1:500, not 1:326!).

Need for calculations

In many cases, the measurement must be used in a calculation before it provides the information required. For example, an area is calculated by multiplication of measured distances. When levelling, the rise or fall of the ground is calculated from the difference between successive measurements. Miscalculations are a common source of error, and need to be double checked before finalising plans.

In many cases, it is not easily possible to measure a particular feature directly; for example, the height of a tower. In these cases, other measurements can be made, from which the desired measurement can be deduced.

TRIANGULATION

Triangulation is particularly useful in surveying. If the required measurement of a feature can be "located" in a right angle triangle, then measurement of any two other parts of the triangle (angles and/or distances) can be used to calculate the required measurement. The rules of right angle triangles are summarised below:

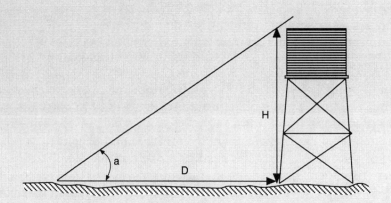

Figure 3.1 *In this example, if vertical angle "a" and horizontal distance "D" can be measured. Then the height "H" can be calculated, using the rules of triangulation.*

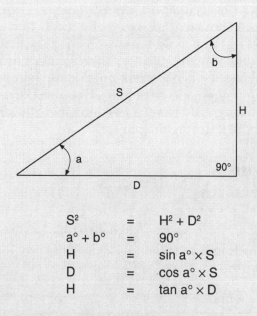

$$S^2 = H^2 + D^2$$
$$a° + b° = 90°$$
$$H = \sin a° \times S$$
$$D = \cos a° \times S$$
$$H = \tan a° \times D$$

Figure 3.2 *The geometric rules of right angle triangles.*

Accuracy and errors

Different tasks require different levels of accuracy; the difference in scale between a site plan and a building plan is an example. It is safest to use a higher level of accuracy than the minimum required. This ensures that rounding off errors do not accumulate. If you are 20 millimetres out when taking each of ten measurements, then it is possible to be 200 millimetres out overall. If you are only 2 millimetres out to start with, then the maximum possible error can only be 20 millimetres.

The most common sources of error are:
- not using or setting up the instrument correctly
- bumping or moving an instrument after it has been set up
- taking a correct reading, but writing it in your notebook incorrectly
- error in calculation
- trying to take a measurement beyond the range of the instrument
- using a damaged instrument (such as a tape measure that is stretched or kinked).

A number of procedures can be employed to minimise the chance of errors:
- Assume that errors are likely. There are techniques available to check if an error has been made, and you should use these before moving off the site. It will be a nuisance to have to return at a later stage and start again.
- Take due care in setting up, taking and recording readings. Double check as required.
- Frequently check the accuracy of all instruments. There are routine tests that can be conducted on surveyor's levels to check they are truly reading level. You could have a "good" tape that you do not use in the field, but only use it to check the accuracy of other tapes.
- Establish a reference or bench mark outside the immediate project area, and take measurements back to it as required. If you know the reference mark is correctly located (position and elevation) you can cross-check back to it during the project.

Measurement of distances
Instruments used

There are many different instruments available for measuring distance. When considering which to use, take into account the accuracy they provide and their ease of use.

Measuring or distance wheel
A measuring wheel is a wheel with a one metre circumference attached to a handle, and fitted with a revolution counter. As the wheel is pushed over

the ground the revolutions, and therefore, metres, are tallied. It is suitable only for applications where readings to 0.1 metres are sufficient. Because the wheel follows the ground, it will not be a true measurement of horizontal distance, as it includes bumps and hollows in rough country thus tending to exaggerate the measurement. In very bumpy conditions, the wheel tends to bounce, thereby further detracting from accuracy.

Tapes
Plastic tapes up to 50 metres, divided into centimetres are commonly used for accurate site measurements. Steel tapes up to 8 metres, divided into millimetres, are used during construction. Be careful not to stretch plastic (or linen) tapes, or kink steel tapes. Treat all tapes with care; wipe dirt and moisture from them as they are retracted, and do not let them snap back when retracting, to avoid damaging the end.

Figure 3.3 *Using the distance wheel*

Chains and bands
When the "chain" (22 yards) was a standard measurement of length, a steel chain consisting of 100 elongated links was used for measuring distance. Metric steel bands (long steel strips 100 metres long and marked with 1 metre divisions, wound on a reel) are occasionally used, but most often as references for calibrating other tapes. Both have largely been replaced in the field for engineering projects by tapes.

Optical distance measurement (tacheometry)
Some surveyor's levels have special stadia lines etched onto the lens, so that staff readings can be used to measure horizontal distances. This feature

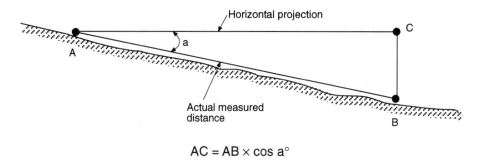

$AC = AB \times \cos a°$

Figure 3.4 *Slope correction using the principles of triangulation*

is useful for site measurements to 0.1 metres horizontal accuracy, because distance and elevation can be read from the one instrument.

Electronic distance measurement
More sophisticated devices are now used by surveyors which provide extreme accuracy, yet are simple and quick to operate. These measure distance by utilising the time taken for electromagnetic or optical waves to travel to a reflector and return to the instrument.

Corrections for slope
Tapes measure along the slope of the ground, but maps and plans require the horizontal projection of the ground to be recorded. If the slope is greater than about ten degrees, a correction needs to be applied to the tape reading, to give the horizontal projection. This is calculated by the rules of triangulation (see page 54 and Figure 3.4).

Offset site surveys

A simple form of site measurement for production of plans and site maps is an offset survey technique, which requires only the measurement of distances and the establishment of right angles.

In Figure 3.5 (on the next page) a base line AB is established on the ground in a suitable location. The positions of important features are located relative to the base line by measuring their right angle distance from the base line, and their distance from one end of the base line.

Sufficient information is recorded to plot these points to scale later in the office, as well as additional measurements as a cross check for accuracy (for example, the length of the sides of the shed). Curves are more difficult to measure, but it is possible by locating one or more points on the curve and the radius of curvature.

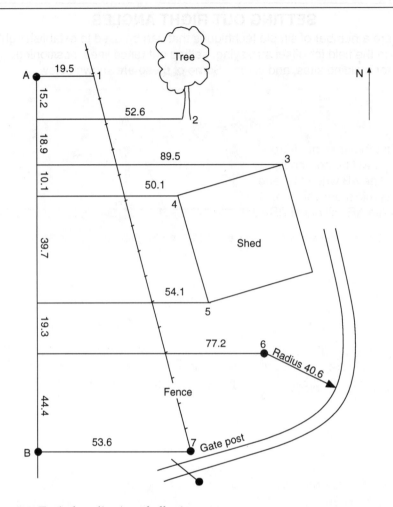

Figure 3.5 *Typical application of offset survey*

Table 3.1 *An example of an offset survey record sheet*

Point	Distance along AB	Cumulative distance along AB from A	Distance from AB	Remarks
1	0	0	19.5	Fence at A
2	15.2	15.2	52.6	Base of tree
3	18.9	34.1	89.5	NE corner of shed
4	10.1	44.2	50.1	NW corner of shed
5	39.7	83.9	54.1	SW corner of shed
6	19.3	103.2	77.2	Centre of curvature of road, radius of curvature 40.6
7	44.4	147.6	53.6	Gate post at B

SETTING OUT RIGHT ANGLES

There are a number of simple techniques that can be used to establish right angles in the field for offset surveying, setting out fence lines, positioning marks for building sites, and so on. Some of these are shown below:

Swing a tape in an arc from point X. It will be perpendicular (90°) to line AB when the tape reads its minimum value, at R. The length AR will equal BR.

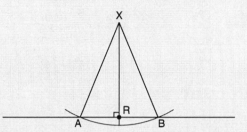

Any point C is marked away from the line AB, and the mid-point of the line AC is located at X. An arc swung at X will intersect the line at B, such that BC is perpendicular to AB.

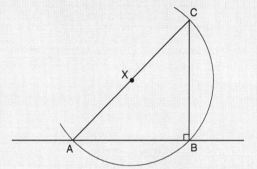

A triangle of the proportion 3:4:5 will be a right angled triangle

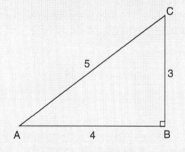

Establish the mid-point X between A and B. Swing an arc through A and B. The intersection of the two arcs is at R, so that RX is perpendicular to AB.

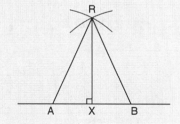

There are also instruments that can be used to establish right angles in the field. One of these is an optical square, which consists of a number of prisms arranged so that by looking through the instrument eyepiece, you can see a line of sight perpendicular to your line of vision on both sides, as well as straight ahead. The line of sight straight ahead is at right angles to the image seen through the prisms.

Measurement of angles

Instruments used

Compass
Compass bearings give sufficient accuracy for larger scale mapping exercises, such as establishing directions and angles between fence lines. Compasses give a bearing of the line of sight relative to magnetic north, not to "true north" used on maps. There should be a conversion factor printed on good maps. Regardless, the difference between bearings in the field will be the angle between lines of sight. Magnetic interference to the compass will cause errors in the readings.

Horizontal circle on optical instruments
Surveyor's levels are made with a 360° marked plate as part of the base of the instrument, which is set up horizontally. Thus, when looking through the instrument's eyepiece, the direction of that line of sight can be noted.
This direction is not a compass bearing unless the circle readings are aligned to north for the initial sighting. However the angle between different lines of sight can still be measured. A theodolite measures both vertical and horizontal angles.

Clinometer
This is a pocket sized instrument consisting of a sight tube, spirit bubble and vertical circle. It is used to measure vertical angles. Whilst looking through the instrument at an object, the circle is turned until the spirit bubble is centred. The inclination is then read off the circle.

Measurement of elevation

It is necessary in construction projects to determine the elevation of the ground surface. During construction, checks for level need to be made periodically. The best method is to use a surveyor's level, which establishes a horizontal line of sight through the instrument from which differences in elevation can be measured.
Small scale maps, such as topographic maps, have elevations measured relative to mean sea level, referred to as Australian Height Datum (AHD).

Figure 3.7 *Using a clinometer*

For most construction projects, it is not necessary to measure elevation to AHD. Rather, a semi-permanent marker is located near the project, but outside the immediate construction and traffic zone. Its position is noted, and it is assigned an arbitrary elevation; 10.000 or 100.000 metres is common. Elevations on the site are then measured relative to this datum.

Optical level

An optical level is an instrument that enables a truly horizontal line of sight to be established. The elevation of objects can be measured relative to this horizontal line of sight, so that differences in elevation between objects can be determined. If the objects are to be at the same elevation, then their elevations relative to the horizontal line of sight must be equal.

Most levels have a number of features:

- Telescopic lenses which magnify the image (like one half of binoculars).
- Crossed lines or similar etchings on the lens (cross-hairs), which enable pinpointing the target exactly.
- A stationary base plate which enables the instrument to be mounted on a tripod.
- A spirit bubble and adjusting screws to enable the base plate of the instrument to be mounted exactly level.
- The facility to rotate the telescope through 360°. Some have a horizontal circle to enable directions to be noted.
- A focus adjustment.

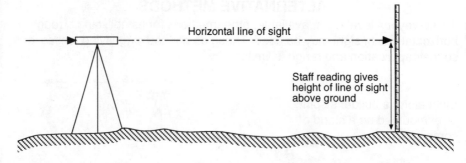

Figure 3.8 *The surveyor's level enables measurements of elevation relative to a horizontal line of sight*

There are a few different types of level. A dumpy level has its telescope fixed to the base plate in such a way that there is no vertical adjustment of the telescope relative to the base plate (the telescope is levelled only by adjusting the base plate). Each time a new object is sighted, the whole instrument needs to be re-adjusted. This type is simple, robust, inexpensive and suitable for building work.

A tilting level has very precise vertical adjustment of the telescope relative to the base plate, and is therefore far more accurate (and expensive). It is also quicker because only one adjustment is necessary, since the telescope can be rotated to view any target.

An automatic level has some lenses mounted on pendulums, so that once the base plate is levelled, further adjustment is unnecessary, making it easier and quicker to use. Of the three types, only the automatic level gives the image the right way up (the image is upside down in the other two).

Figure 3.9 *Typical surveyor's level (automatic type)*

Site measurements and plans

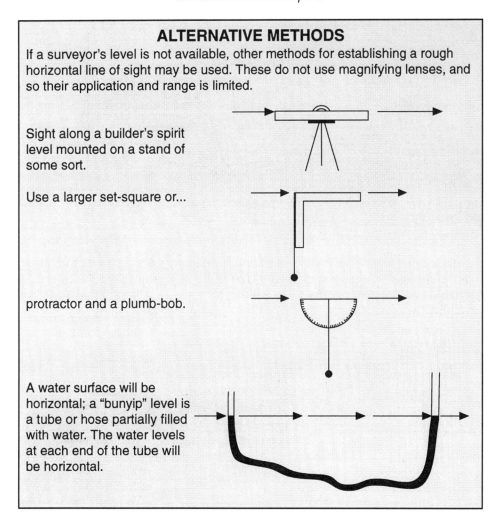

ALTERNATIVE METHODS

If a surveyor's level is not available, other methods for establishing a rough horizontal line of sight may be used. These do not use magnifying lenses, and so their application and range is limited.

Sight along a builder's spirit level mounted on a stand of some sort.

Use a larger set-square or...

protractor and a plumb-bob.

A water surface will be horizontal; a "bunyip" level is a tube or hose partially filled with water. The water levels at each end of the tube will be horizontal.

Setting up surveyor's levels

1. Extend the tripod legs and establish the tripod position. Make sure the legs of the tripod are firmly set into the ground, and all tripod connections are tight. Any movement after you start taking readings will cause errors. The tripod top plate needs to be approximately level (by eye will do).
2. Mount the level firmly on the tripod using the central attaching screw. Do not overtighten.
3. The base plate of the instrument is levelled by means of the three levelling adjustment screws. Turn the telescope so that it is parallel with any two levelling screws (for example A and B in Figure 3.10). Rotate A and B only until the spirit bubble moves to the centre of its travel in the direction of the telescope. When rotating A and B at this stage, turn them in opposite directions only. Then rotate adjusting screw C to centre the spirit bubble. Lastly, rotate the telescope to a number of different positions, using the levelling screws for fine adjustment.

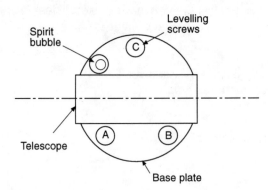

Figure 3.10 *Levelling the base plate (looking down on the instrument)*

This procedure levels the base plate of the instrument only, regardless of the type of level. If you are using a dumpy or automatic level, they are now ready for use. For any level, do not bump or lean on the tripod once you have got this far!

In the case of a tilting level, there is an additional adjusting screw under the telescope itself, and an additional spirit bubble on the telescope. For every sight of the instrument, fine adjustment of this screw is required.

4. Turn the eyepiece of the telescope so that the cross-hairs are focused.
5. Rotate the telescope towards the object. There are usually a couple of sight bumps on the top of the telescope to help initially, and often a fine adjusting knob.
6. Focus on the object and take a reading.

Reading the staff

Now that a horizontal line of sight has been established, the height of the ground relative to it needs to be measured. This is normally done with a surveyor's staff, which is an extendable aluminium staff usually marked in centimetres.

Some points to remember:

- The staff needs to be held vertically. Good concentration helps, as does a builder's spirit level.
- Do not try and read the staff if it is either too far away or too close.
- Record the position of the staff in some way, using a peg or other markers. Beware of livestock (and human) interference to the peg.
- Position the staff on a representative position, not on top of a big clump of grass or in an obvious but localised depression such as a wheel track.
- Get used to reading the staff upside down, because the image is inverted unless you are using an automatic level.

Figure 3.11 *Surveyor's assistant holds staff at location where relative elevation is to be measured. Note staff graduations are in centimetres*

Permanent adjustments and a test for accuracy

There are adjustments that can be made to an instrument to ensure it does give a truly horizontal line of sight when the spirit bubble says it does. These are set by the manufacturer and should not be adjusted unless you know what you are doing.

On the other hand, it is important to know whether the instrument is reading accurately or not, and if not, by how much it deviates. The "two peg test" is a simple and fast method for testing a level, and would be conducted when using a level for the first time, or after suspecting the level requires adjustment.

Two pegs (A and B, see Figure 3.12), are located about 60 metres apart. Staff readings are taken to both pegs, from two staff stations, one at the midpoint between the pegs (X) and one from just outside one of the pegs (Y). In the sketch in Figure 3.12, the staff readings are labelled 1, 2, 3 and 4.

If the level is accurate, then $3 - 1 = 4 - 2$. If not, then the instrument needs re-adjustment.

Standard plan presentation

Figures 3.13 and 3.14 show extracts from a typical plan for a farm building. The following features should be noted:
1. A description of the drawing is provided, usually in the lower right corner. The project name or owner should be included along with the name

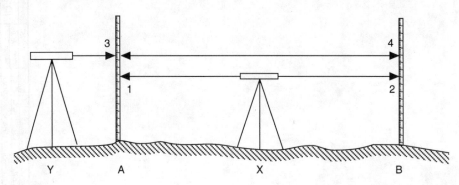

Figure 3.12 *Two peg test*

of the person or organisation that drew the plan and the date it was drawn, and a drawing/plan number if it is one of a set. The scale should be noted here if all drawings on the plan are drawn to the same scale, otherwise the scale is shown under each drawing. NTS means "not to scale", in which case all dimensions should be shown.

2. The one sheet can show multiple drawings; in fact, it is preferable to have all the drawings together this way. For a building, there would be a site map, a plan of the building (ie when looking down with the roof off), and a number of elevation drawings (to show the side-on view from various directions). It would be common to show a cross-section through the building, to show the typical construction detail. Additional detail drawings would be provided as required, for more complex constructions (such as Figure 3.14).
3. The drawings show the details of the below ground footings, not just above ground parts of the construction. For kit sheds the standard plans may not show footing details because they may be site specific, and so would say "to engineer's specifications".
4. Although drawn to scale, important dimensions are noted on the drawings, in millimetres for building plans.
5. Important specifications can be shown on the plan, in addition to a full list of specifications and quantity of materials that would be compiled separately, and forming part of the contract documents. For example the label M16 on some of the bolts shown in Figure 3.14 describes them as 16 millimetre mild steel bolts.

Pegging

Once measurements and calculations such as for earthworks have been made, it is still necessary to instruct the operators of the earthmoving machinery where to move earth from, where to put it, and how much in each

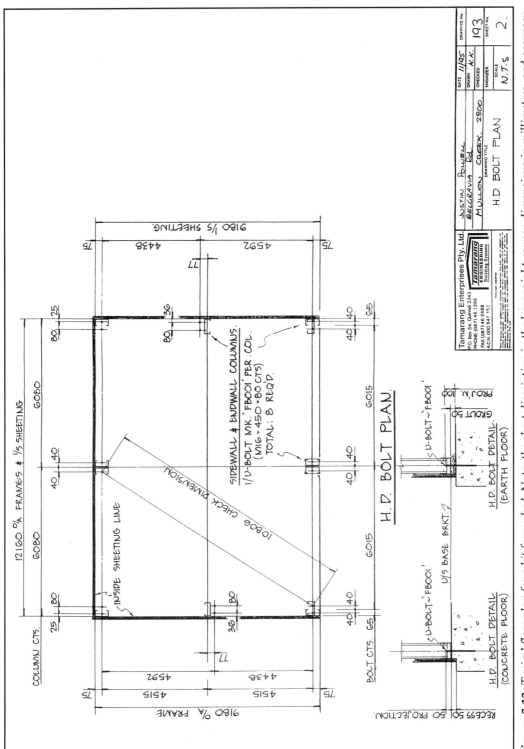

Figure 3.13 *Typical floor plan for a kit farm shed. Note: the sheet discription in the lower right corner, dimensions in millimetres, and accuracy is to the nearest millimetre The abbreviations O/A, I/S and CTS refer to overall, inside, and centres. The details for holding down (H/D) bolts are shown for two possible floor options, but the actual size of the footings is listed elsewhere. (Courtesy Tamarang Enterprised Pty Ltd.)*

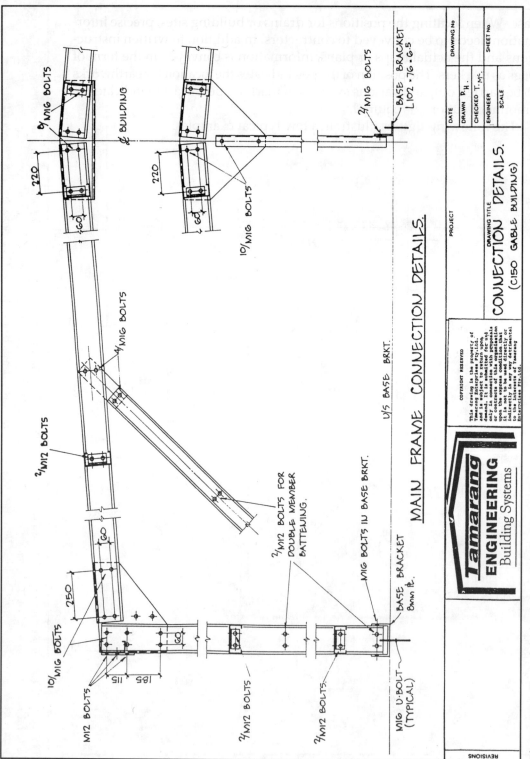

Figure 3.14 Typical drawing for construction detail. Some important specifications are also included, for example 10/M16 bolts refers to 16mm diameter mild steel bolts (10 of them). PL stands for plate. Dimensions in millimetres. (Courtesy Tamarang Enterprises Pty Ltd.)

case. When locating the positions for drains or building sites, precise information needs to be conveyed to contractors. In addition to written instructions and the actual maps or plans, information is conveyed in the form of pegs as markers. The position of the peg indicates the location for earthworks to be carried out, and various systems of marking pegs exist to indicate the amount of cut or fill required.

The following figures illustrate a few typical examples:

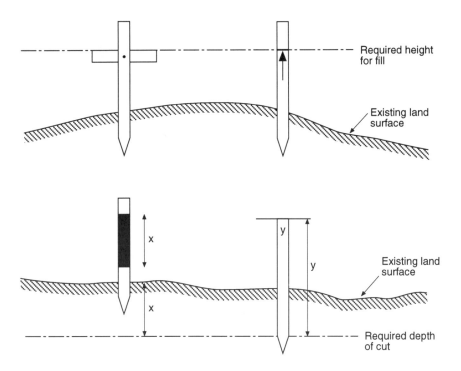

Figures 3.15 *Pegs for setting out amounts to fill or cut at the location of the peg. The peg is kept in position until the earthmoving is approaching completion*

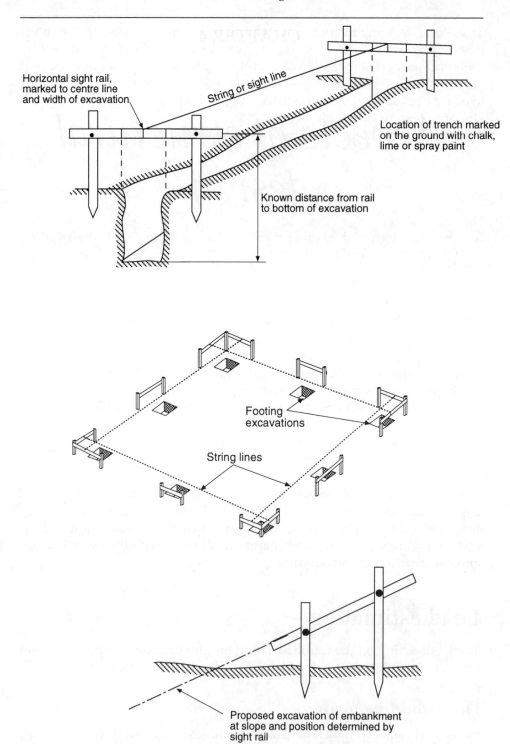

Figures 3.16 *Sight rails; usually for setting out lines but also for systems of lines, the intersection of which indicates the location of an important feature*

CHAPTER 4

Aspects of structural design

This and the next chapter describe some aspects associated with the structural design of farm buildings. They are not meant to convert the reader into a structural engineer overnight, but are included to illustrate the procedures involved in building design, and to enable better assessment of alternative construction methods. It should help when trying to select a design for a new building, by identifying the key factors which determine its structural integrity. It should also help identify the best method of repairing structural defects in older buildings.

This chapter focuses more on the loads applied to farm buildings, and how individual components of the structure are designed to withstand the stresses, strains and deflections which those loads create. The design must cope, to a predetermined level, with normal and extreme loads, including those caused by severe storms. This applies during construction and not just after completion. The next chapter deals more with the overall design and construction of the building.

Load estimation

The loads acting on the frame of a farm building can be grouped into three main categories; static or dead loads, live loads and wind loads.

Dead and live loads

Dead load refers to the static weight of the structure itself. Live loads are those that fluctuate during the life of the building, and include loads exerted by stored commodities or suspended equipment and materials. Floors and rooves are subject to live loads because of the potential for traffic, including

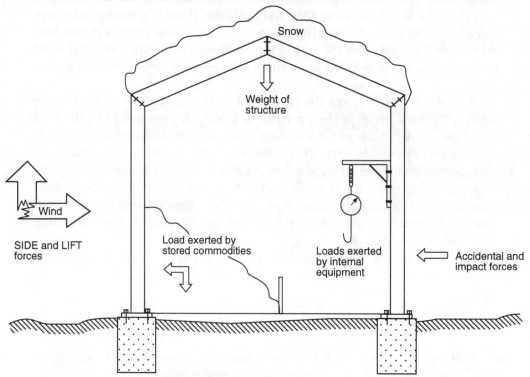

Figure 4.1 *Any building is subject to a variety of loads, each of which must be considered in the structural design of the building*

such things as occasional walking on the roof by people during construction and repair. Even the duration of the loading needs to be considered, and the effects of expansion and contraction due to temperature fluctuation. For farm buildings, where the floor is normally either the ground or a concrete slab on the ground, and where roof live loads are unlikely, relatively simple structures are possible. The exceptions are those buildings which contain a raised floor, such as shearing sheds, or storage sheds with an elevated loading dock, where more detailed consideration of loadings is required.

Sheds, silos and tanks required for storage of bulk grain and liquids also require detailed consideration of load conditions, both live and dead. A substantial outwards pressure is exerted by bulk grain, fertiliser and liquid in storage, and the structures used for these purposes must be designed accordingly.

Wind load

Dead and live loads can be estimated reasonably accurately. The force of the wind on a building is the most severe load condition, and one which is more difficult to predict with certainty. The Loading Code for wind forces provides specific guidelines to be followed.

When the wind blows against the building, there is a strong sideways force trying to push the building over. The size of this force is dependent on the wind velocity and the area of the building's wall exposed to the wind (a small increase in wind speed causes a large increase in the force generated).

As the wind blows over the roof of the building, it creates a suction effect on the roof because the pressure is lower on the outside of the roof compared to the inside of the building. This effect is worse when the building has one open side exposed to the wind and walls on the other three sides, pressurising the air inside. The effect is less severe if internal air pressure can escape through door or window openings, where the wall cladding does not extend to ground level, or where the building has less than three walls with cladding. A fully enclosed building is less affected than a three sided building.

The combined effect of these forces is to try and lift the building off the ground. In fact, a lot of effort goes into holding the building down, as well as holding the building up.

The force of the wind can lift roof cladding off, but if the cladding is securely fastened to the roof frames it has the potential to lift the whole roof off the walls, or even lift the whole building off the ground. The building is anchored to the ground by its footings, and each part of the frame and cladding must be secured in a way which prevents it lifting.

Figure 4.2 *The damage to the round yard at this horse education centre was caused by part of the skillion roof from the far side of the other shed being lifted during a storm, and falling onto the yard in one piece*

Figure 4.3 *This three sided building was severely damaged during strong wind. The rear part of the roof has been lifted off, and there has been separation of the point of connection between the centre roof truss and column*

Figure 4.4 *The same site (Figure 4.3), from downwind. Roof cladding is strewn on the ground, and wrapped around tree branches*

Estimation procedure

The task of the structural engineer is to predict the likely magnitude of the loads on the building. The appropriate Loading Code must be consulted. For most farm buildings, dead and live loads are relatively easy to predict. For wind loads, the Loading Code nominates a basic wind speed that is likely for the area where the building is to be located.

This basic wind speed needs to be modified according to a variety of factors, including the local terrain and any shielding effects, since the topography of the immediate vicinity can influence wind conditions. For example, a building designed to withstand likely wind in terrain Category 2 must be more strongly built than a building designed for terrain Category 3. (There are other Categories. You should check on which category is appropriate before accepting design.)

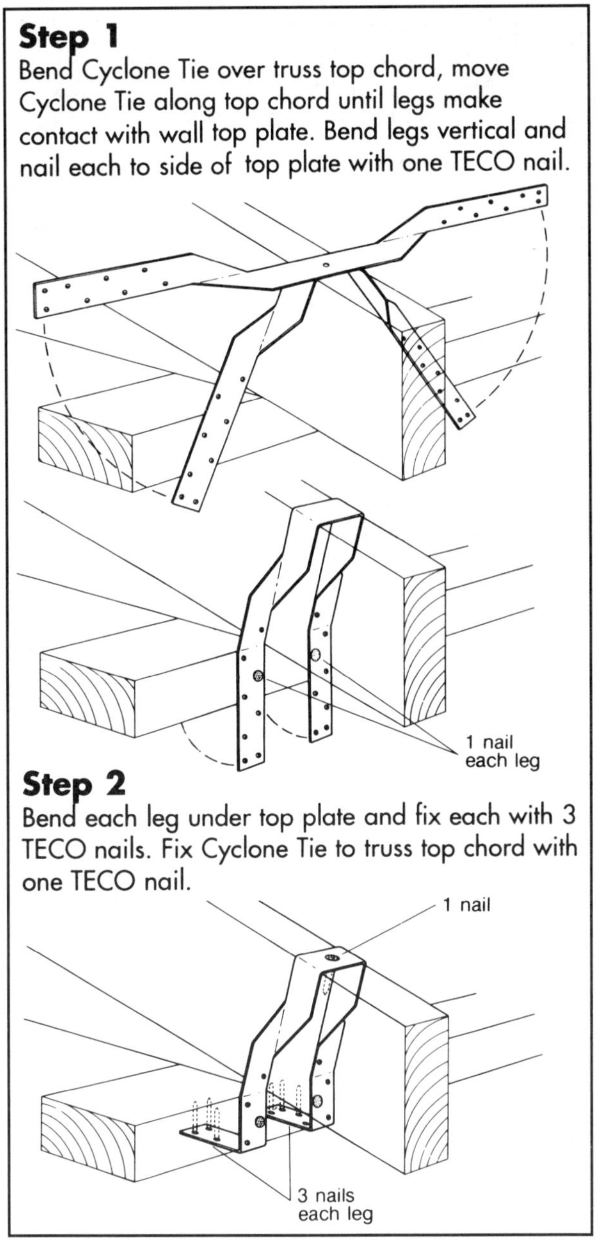

Step 1
Bend Cyclone Tie over truss top chord, move Cyclone Tie along top chord until legs make contact with wall top plate. Bend legs vertical and nail each to side of top plate with one TECO nail.

1 nail each leg

Step 2
Bend each leg under top plate and fix each with 3 TECO nails. Fix Cyclone Tie to truss top chord with one TECO nail.

1 nail

3 nails each leg

Figure 4.5 *Installation instructions for a Teco cyclone strap which will help hold a timber framed roof against strong wind. (Courtesy Gang-Nail Australia Pty Ltd.)*

Even so, the basic wind speed is a nominal one. It is based on records of wind speed across the country, but it is not the highest wind speed ever recorded. It has a mean return period of 50 years (that is, it is likely to occur on average, once every 50 years). Farm buildings, with a low hazard to life and property if they fail and fall down, can be designed with a slightly lower design wind speed (a mean return period of 25 years). This makes farm buildings less expensive to build, but also means they are more liable to be damaged in high wind situations. Take advantage of local terrain and shelter to help reduce the likelihood of wind damage. Note that buildings in a specifically designated cyclonic wind area (coastal areas north of 27°S latitude) must be constructed to a different set of conditions again.

Having decided the forces the building is likely to encounter, the structural engineer then must calculate how strong to make the component parts of the building, and how they are to be fixed together.

Design of structural components

The forces generated by the imposed loads create stresses in the components that make up the building. Strength and stiffness are two important characteristics required in building components to resist the loads which are acting on them. The strength of each component must be substantially greater than the stress within it, otherwise it risks failure. Strength is determined partly by the characteristics of the material the component is made from (for instance steel is stronger than timber, hardwood is stronger than softwood), the size of the cross-section of the component, and the way the material is distributed across the cross-section.

As an illustration to these factors, consider a beam supported at both ends as in Figure 4.6.

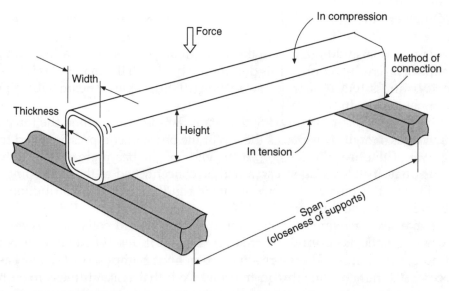

Figure 4.6 *A beam supported at both ends; various factors determine its ability to withstand the loads applied*

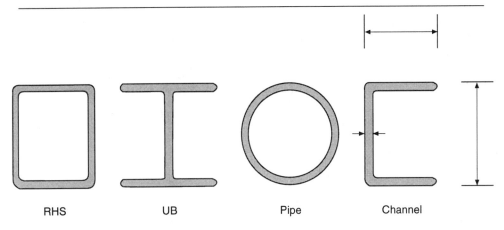

Figure 4.7 *The shape of the cross-section will influence structural characteristics, as well as height, width and thickness*

The maximum vertical force acting on the beam has been estimated to be a certain amount, based on the likely loads acting on the structure and how the loads are transmitted to this component. The strength of the beam (its ability to resist the forces acting on it) will be proportional to its:

- height squared
- the square root of the span
- width
- characteristic strength of the material the beam is made from, its wall thickness and its freedom from defects, and
- whether the ends of the beam are fixed rigidly to the supports (welded), or free to move a little when the load is applied (bolted).

Such calculations enable the engineer to determine what size beam is necessary.

There are also differences in the strength of the beam depending on the shape of its cross-section as this determines how well the material is located to resist stress; rectangular hollow section (RHS), universal beam (UB), pipe, channel and so on (Figure 4.7).

In the example in Figure 4.8, the beam risks bending when the force is applied as shown. This places the top of the beam in compression, and the bottom of the beam in tension. If the ends of the beam are able to move somewhat, the resulting stresses will be concentrated at the mid-point of the beam. If the ends of the beam were rigidly fixed, the distribution of stresses would be different.

If the stresses were high enough, the beam would collapse. However, excessive deflection of the beam, even though it doesn't collapse, is also considered a failure. Therefore a deflection limit is imposed, and the beam must be designed with this in mind as well; that is its stiffness must be sufficient. Strength and stiffness can be influenced by the inclusion of bracing. This is common in building design, particularly farm building design,

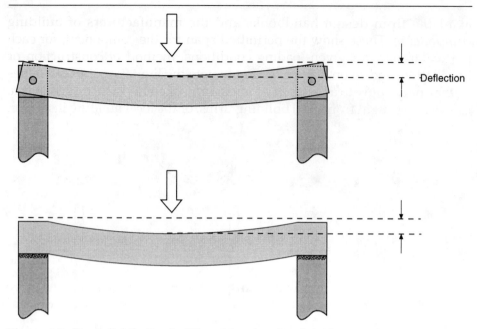

Figure 4.8 *Stress distribution is different for pinned and rigid connections*

because the inclusion of simple bracing components can greatly reduce the cost of the main frame components.

Even if the beam is designed with great strength, it can still fail by coming apart from its supports. Consequently, the design of the connections between components needs just as much consideration as the components themselves. For example, a bolted connection can fail by the nut being pulled off the bolt, by the bolt pulling through the bolt hole, or the bolt shearing apart. It follows that a bolted connection also needs to be designed correctly; the correct number and diameter of bolts tightened to the correct tension, bolt holes of the correct diameter, and with the material properties of the components in mind. Similar considerations apply to welded connections.

The forces on the beams are transferred to its supports, and their design is subject to detailed investigation. In addition to the above, a column subject to vertical compression risks buckling, and when subject to sideways forces, excessive bending and deflection. Its connection to the building's footings must be secure.

Any twisting forces applied to the building components create torsional stresses, and these too need to be checked.

The above process highlights the complexity of structural design. However, there are some points to consider that make construction projects easier. When purchasing a kit shed, the manufacturer will have plans and specifications on hand prepared by a structural engineer, which are suitable for submission direct to the approving authorities. Span tables are also available from design handbooks and the manufacturers of building components. These show the permitted span for the component, for each particular size, for given loading conditions, method of construction or installation, and so on. It is important to note the limitations of span tables — they only apply to a particular combination of circumstances. Assistance can also be sought from local building firms, or the local building inspector.

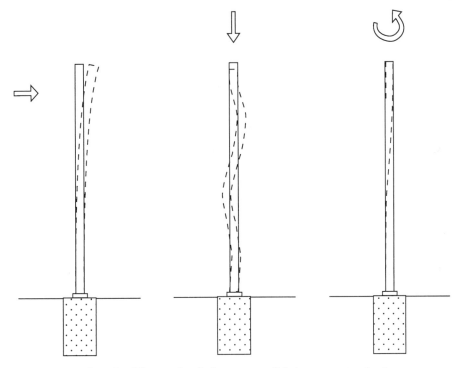

Figure 4.9 *Bending, buckling and twisting are possible in a structural column*

CHAPTER 5

Types of construction

There are a number of different types of construction for farm buildings. The most popular of these is portal frame. Each type achieves strength, stiffness and stability by different methods. This may influence the cost of the building, and how well it meets its design objectives.

Portal frame construction

Portal frame construction, using all steel components, is in common use for rural and industrial buildings. This method is relatively inexpensive, fast to erect, adaptable to a wide range of spans and lengths, gives a neat and trim appearance, and provides clear headroom under the roof. Alternative construction materials like timber beams and purlins are possible but not common.

Frame details

The name is derived from the portal frame that spans across the shed. This frame consists of two columns and two roof beams, with rigid connections at all three points of the frame. The frame is then secured to substantial footings by holding down bolts. The size of the frame components and the rigidity of their connection ensures there can be no movement or deflection across the frame. Adjacent frames are connected by purlins and girts, using as many frames as is required to give the required length of building. The end walls are attached with their own frame of columns and girts.

Although each frame is itself secure against lateral forces, there is not much support against forces in the longitudinal direction. The holding down bolts are not designed to cope with such extra load. Consequently the frame

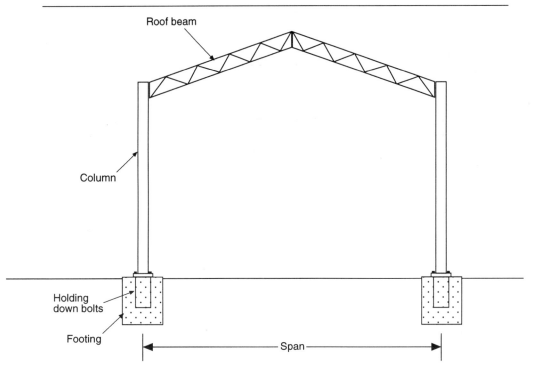

Figure 5.1 *Portal frame*

requires cross-bracing to stiffen it in the longitudinal direction. This is located in the walls and both sides of the roof. If one bay is braced, adjoining bays can not move because of the connection of the purlins and girts. However, because of the clearance between the purlin bolt and their bolt holes, long buildings will require bracing at each end bay, and possibly in the centre bay.

There are a few variations of detail for this type of building frame. Knee and apex braces or gussets are commonly used in the three points of frame connection, to assist in stiffening them. This reduces reliance on any single bolted connection (even though multiple bolts are used) because of the triangulation provided by the brace. Occasionally, the frame is welded rather than bolted, but this is not common. Frame components can be made in heavy duty materials, which impart strength and stiffness due to the inherent strength of the material and its cross-section. The alternative, common with kit sheds due to the freight advantage, is to use light steel sections, but use additional bracing to provide stiffness; for instance cross-bracing in all roof bays and the end walls.

Multiple spans are possible to economically extend the area covered under one roof. This technique keeps the frame cost down, compared to a single span frame of the same span, but requires intermediate columns that may interfere with the internal use of the building. A further complication is the need for box guttering where roof spans meet.

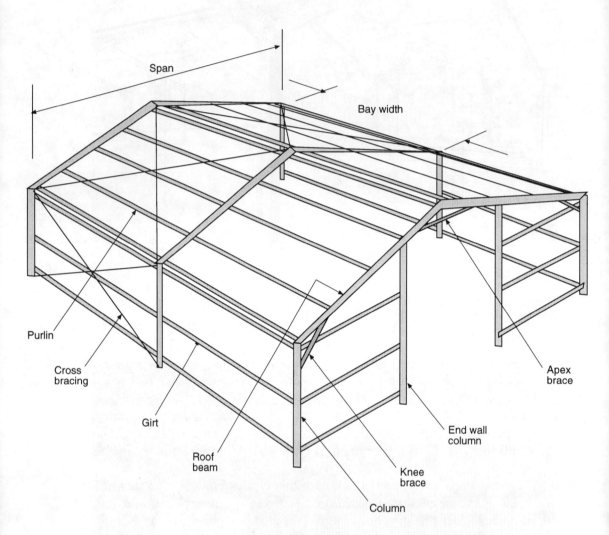

Figure 5.2 *Schematic representation of main frame components for portal frame construction. For a given frame span, multiple bays are added to give the required building size*

Most buildings of this type would use steel C-section purlins (for the roof) and girts (for the walls). They are inexpensive, strong yet lightweight, and can span further than timber. The purlins are bolted end to end onto a cleat welded or bolted to the roof beams, (in the case of purlins), and columns, (in the case of girts).

A standard span for the purlins, and therefore the spacing between frames, is six metres, but variations are possible. Some farm buildings may need to have at least one bay wider than this to allow entry of wide equipment. Purlins are available with bolt holes pre-punched at their ends.

Types of construction

Figure 5.3 *Interior view of a multiple span portal frame greenhouse, showing minimum interference to growing area*

Figure 5.4 *Portal frame machinery shed, in this case with a welded main frame, and steel purlins and girts. Note the cross-bracing in wall and roof of end bay, and inclusion of stiffening struts at the mid-span of the purlins*

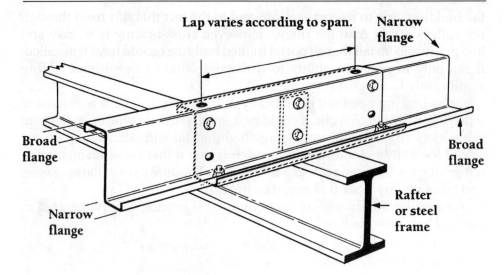

Figure 5.5 Z-section purlins can be overlapped, which adds strength. (Courtesy of BHP Building Products)

Variations in purlins are possible. Z-section purlins are available, where it is necessary or desirable to overlap the ends of the purlins; one nests inside the other for a small distance. C-section purlins can not do this. The centre of the overlap bolts to the frame cleat, and the ends of the overlap are also bolted. A certain length of overlap is required, but it effectively creates a continuous purlin for the full length of the building. This provides greater strength, so it can span further, but the length required is specially selected for the job.

The load carrying ability of a purlin can be increased by inclusion of purlin stiffeners (or bridging). There are simple bracing rods or struts which stop the inside edge of the purlin buckling when a load is applied. (The cladding stiffens the outside edge.) They are usually bolted or clipped into position, bridging the space between purlins. In some designs, they are essential rather than optional depending on the load and span for that design. Sometimes two stiffeners are required.

Timber purlins are possible, but heavy sizes are required to span the same distance as steel. They are also more subject to warping, sagging and deterioration from the weather.

Cladding is screwed directly into the purlins and girts. Nails are possible for timber purlins and girts, but screws are a far superior method of fixing. (See Cladding in Chapter 6.)

Cross-bracing

This is an integral part of the portal frame design. In some buildings the design does not rely on cross-bracing, rather the cladding on the walls stiffens the building, due to its corrugations and to the fact that it is fixed through the valleys rather than the ridges. However, cross-bracing is so easy and inexpensive to install that all portal framed buildings could have it installed; it can only improve its ability to withstand wind loads without adding significantly to the cost.

Cross-bracing need not be of heavy duty material, since it is in tension when it is required to work. It could be a small diameter pipe or angle, wire cable, steel strap, even heavy gauge high tensile wire, depending on the size of the building and its wind load. It is best if there is some method of tensioning the bracing, such as a turnbuckle. This makes installation easier, and takes up any slack that might be present.

For smaller buildings (say up to four bays) bracing is only required in one end bay, provided:

- the walls and roof of that bay are braced
- the bracing is crossed
- it is attached to occupy the full area of the bay as shown in Figure 5.6.

In longer buildings bracing may be required in each end bay.

Occasionally this may interfere with proposed door or window openings, but do not compromise the integrity of the frame. Either relocate doors and windows, or relocate the bracing into other bays. This is best considered at the planning stage.

Sometimes cross-bracing is a single diagonal component, rather than crossed. In this situation a heavy duty cross-section is necessary because the bracing must work in compression as well as tension.

Footings

Footings need to be of sufficient width to prevent sinking into the foundation, and sufficiently heavy to anchor the building down during wind uplift. Follow the manufacturer's or engineer's instructions in this regard. In many

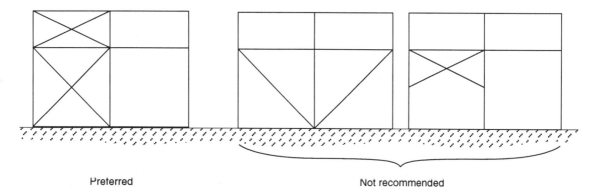

Preferred Not recommended

Figure 5.6 *Preferred bracing arrangement*

Farm Buildings

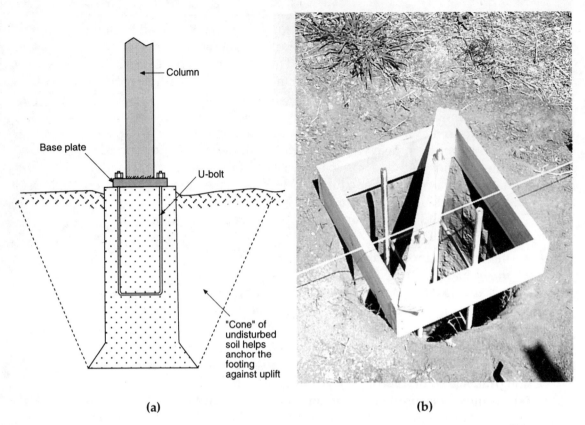

Figure 5.7 (a) *A typical footing arrangement for a portal frame building. Belling the bottom adds greatly to its anchoring in the ground.* (b) *A mass concrete footing with holding down bolts clearly visible to anchor the column*

farm buildings it is sufficient to drill a hole with a 450 millimetre post hole digger, to about 900 millimetres depth. If the bottom of the hole is belled out, greater anchorage is obtained, because the footing is held down by a large mass of soil, in addition to its own weight of concrete. Deep footings should include some reinforcing steel, but ensure it is covered by at least 65 millimetres of concrete. The holding down bolts are embedded deep into the concrete, so they will not pull out of the footing when wind uplift occurs. Use long U-bolts, or weld up a frame that secures them together. Building specifications may require two or four bolts. A base plate attached to the bottom of each column then slips over the bolt and provides the method of securing the frame to the footings. The top of the footing may require stepping to accommodate the bottom edge of the wall cladding. It should be sufficiently above ground level that water does not accumulate around it.

Construction procedure

1. The site is normally levelled, since the column height is normally equal (although unequal column height can be specified if necessary). Provide

Types of construction

Figure 5.8 *Holding down bolts are positioned prior to pouring a pier type footing. Note the string line for alignment*

for stormwater run-off away from the site by construction of interception drains and/or appropriate grading away from the building area.
2. Set up sight rails or other means for marking out the building.
3. Excavate for the footings. When digging out the footings and positioning the holding down bolts take care that they are in exactly the right

Figure 5.9 *Footings for this shed were dug with the post hole digger. Note the protruding holding down bolts*

Figure 5.10 *Main frame during erection*

location; straight square, and at the correct level. If not, the building frame and cladding will be out of square. Re-check immediately after pouring the concrete that the position of the holding down bolts is still correct. Allow time for the concrete to cure.

Figure 5.11 *When the main frame is complete the installation of windows and cladding can commence*

4. Locate and bolt columns to footings. Roof beams can be lifted together after bolting on the ground. Alternatively, the whole frame can be bolted on the ground and lifted up in one piece. Be prepared to use lifting machinery for heavy duty construction.
5. Use temporary bracing to hold the frame in position until it is complete. Loosely bolt connections, starting with apex purlin and top girt. Install end wall frame.
6. Square up the building and install cross-bracing. Re-tighten all bolted connections to correct tension.
7. Install purlin and girt stiffeners.

The structural frame is complete, ready for installation of windows, doors, cladding, gutters and weatherproofing (see Chapter 6).

Kit sheds

Kit sheds are popular for rural applications. They provide a low cost construction that many farmers are able to erect themselves. General purpose storage buildings are readily available in kit form, and some manufacturers

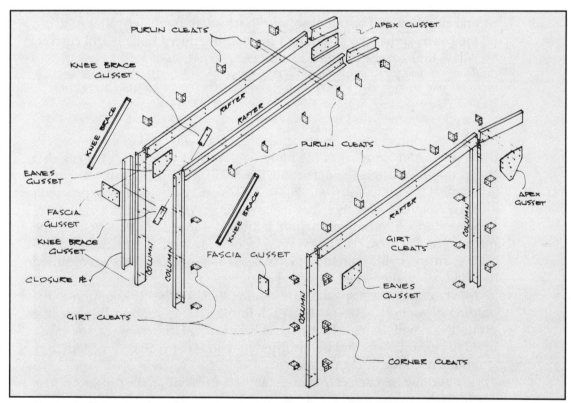

Figure 5.12 *Extract from instructional manual for a typical kit shed, showing main frame parts. Note the use of gusset plates and knee braces to stiffen the frame and facilitate simple bolted connection. (Courtesy Tamarang Enterprises Pty Ltd.)*

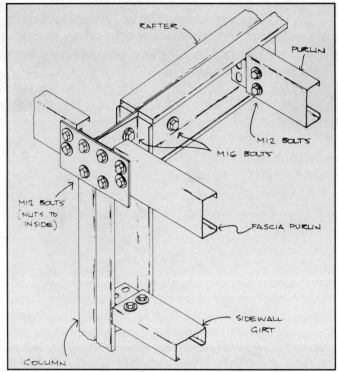

Figure 5.13 *Detail of connection between roof beam (rafter) and column, and connection of purlins and girts. In this design the purlins and girts are bolted to cleats located within the flanges of the rafter and column. This is to reduce wasted wall thickness, and simplify installation of internal wall lining. The bolted connection is stronger than fixing with self-tapping screws. (Courtesy Tamarang Enterprises Pty Ltd.)*

offer special purpose buildings such as shearing sheds. Make sure the overall dimensions and function of the building suit the purpose, and check that the structural design (in particular the wind load it is designed for) is adequate. (The plans shown in Figures 3.13a and b are for a kit type shed.)

Lightweight steel sections are used, relying on extensive bracing to provide support. Alternatives include lattice or webbed construction for the main frame. For farm buildings it may be best to avoid construction materials that are too light; even though marketed as sheds, some are more suited to applications such as garages and carports.

Most designs use a special base bracket that attaches the column to the holding down bolts. This enables each frame to be assembled lying on the ground, and with a single bolt on each base bracket, the frame can be easily lifted to vertical. A second bolt is fitted to hold it there. Such a system may avoid the need for lifting equipment.

To minimise cost, select from the standard sizes within the product range. These also require a level site. Door and window options are quite flexible. Non-standard structural design is more difficult, but negotiate with the manufacturer.

Types of construction

Figure 5.14 *The early stages of kit shed frame erection. Main frames are assembled at ground level, and fixed to propreitary base plates secured to footings. This enables the frame to be lifted easily, pivoting on the base plate. (Courtesy of J. Powell.)*

Figure 5.15 *Construction is now nearing completion. Note the extensive use of cross-bracing. The end wall frame includes provision for windows, but gutters and barge and corner capping are not yet installed. (Courtesy of J. Powell.)*

Pole-in-ground construction

Pole-in-ground type of construction uses a timber pole or steel column buried firmly in the ground to provide structural strength and rigidity. (Steel columns are concreted in.) If the pole is embedded to sufficient depth, and the ground compacted around it, there should be no movement of the pole when loads are applied from any direction. A roof frame, of either a truss design or a simple beam, is attached to the top of the poles. Purlins spanning the trusses or beams enable roof cladding to be attached, and wall cladding is attached to girts spanned between poles.

This type of construction was used in many older farm buildings. It has been largely superseded by portal frame construction for new buildings. Good quality timber poles are scarce and expensive. Steel columns could be used with home made roof trusses, but portal frame construction usually ends up quicker and cheaper anyhow, if all materials have to be purchased. Trusses generally require a bottom chord running between the tops of the columns, possibly restricting roof space for tall machinery or hay storage. Stable ground conditions are essential to ensure the poles stay vertical, although cross-bracing can be added if necessary. Pole framed buildings are sometimes preferred on steeply sloping sites or where a creative design is preferred.

Figure 5.14 *In this older style hay shed, timber poles have been used, with roof trusses made of steel pipe. (The end wall, with its steel frame, is a recent addition). Note that the sapwood has been removed from the base of the pole prior to its erection*

Pole selection and embedment

Timber poles can be used, but because they will be in direct contact with the ground, their durability must be considered. Some timbers are sufficiently durable in well drained, stable soil types (if termites are not present). Timber preservative and soil treatment can be used to protect less durable timber species, or as a general rule wherever timber is used for in-ground applications. Timber poles should not be encased in concrete because as they shrink, a gap will be left around them. If they swell, the concrete can be cracked. Some cement powder mixed with the backfill, however, can be used effectively.

Steel columns have the advantage of availability, but because they are relatively small in cross-section compared to a timber pole, they need embedding in concrete to provide stability.

Pole or column diameter will depend on the height of the pole, the loading conditions it experiences, and the strength of the material in use.

Sufficient depth of embedment is very important, since this has the greatest influence over pole stability. Friction between pole and ground is used to resist uplifting forces. In stable soils under a low wind load, the minimum depth of embedment is around 1.2 metres. As the height and span of the structure increases, the required depth of embedment increases further. This makes it quite a difficult job to install the poles, and increases their total length and cost, and possibly explains why many such buildings end up with a lean. In soft soil, or those subject to periodic saturation, the required depth of embedment makes this method of construction impractical. Close pole spacing is also often required, since more poles are needed to assist stability, but this adds to the cost for a given floor area.

Roof frame

A simple webbed truss can be used to span between poles. Various designs are used. They can be made out of relatively light materials, since the web pattern stiffens the truss, and transfers downward loads to the bottom chord, which will normally be in tension. Alternatively a straight structural beam could be used to span between poles, providing a flat sloping roof.

The truss needs to be secured to the top of the pole to resist wind uplift. When this occurs the bottom chord of the truss is placed in compression, and buckling is a risk. Consequently, long span trusses require bracing in the longitudinal direction through the bottom chords of the trusses. Additional stiffness is achieved by bracing between the apex of one truss and bottom chord of the adjacent one.

Purlins are attached to cleats on the top chord of the truss. Girts are attached directly to the poles.

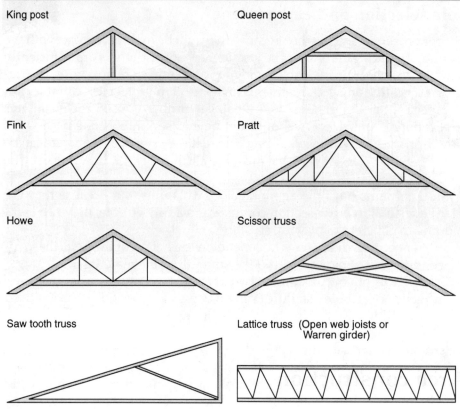

Figure 5.17 *A selection of alternative roof truss designs*

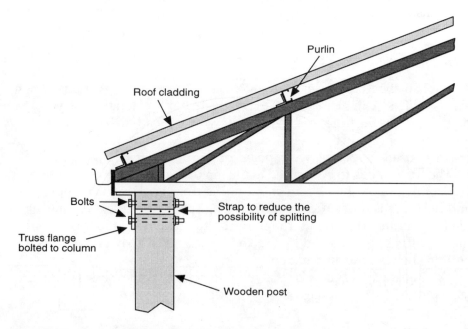

Figure 5.18 *One method of anchoring a truss to the pole*

Figure 5.19 *Truss being lifted into position, ready to be bolted to the column top plate*

Arch construction

Arch construction is also sometimes referred to as "hoop", "tunnel" or "igloo" construction. The structural principle of this design is that radial loads are transferred to the footings due to the continuity and shape of the arch. If

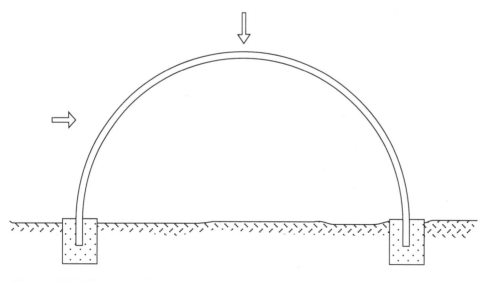

Figure 5.20 *The shape of the arch transfers all loads to the footings*

Figure 5.21 *An arch frame hayshed*

Figure 5.22 *An arch frame greenhouse with plastic film covering. Note the short subwall and provision for cross-flow ventilation*

Figure 5.23 *The raised arch of this piggery building (under construction) provides a reasonably high vertical side wall. This has been braced by an external buttress*

the ends of the arch are embedded in the footings, then some support is also provided to longitudinal loads, although the slenderness of the arch material usually requires bracing. Purlins are fixed directly to the arch, and cladding attached. In greenhouse applications, a plastic sheet forming the skin is simply stretched over the frame.

Arch construction is relatively inexpensive. It is mostly used for greenhouses, but is also found in general purpose buildings. There are two main problems; the curvature near to ground level limits useful floor area, and because the load bearing part of the structure must be continuous across the shape of the arch, ridge ventilation is more difficult.

The first of these problems can be offset by changing the shape of the frame to include a vertical wall or sub-wall, but this compromises the nature of the arch and additional bracing may be required. The second problem is only a problem if ridge ventilation is required. General purpose buildings do not require it. Cross-flow ventilation may be sufficient in single or double span structures. Forced ventilation can be provided, if necessary, with fans.

Concrete slab construction

The properties of concrete and working with it are discussed in Chapter 2.

In portal frame construction, the structural loads from the building are transferred to pad or pier footings under each column. A concrete slab floor

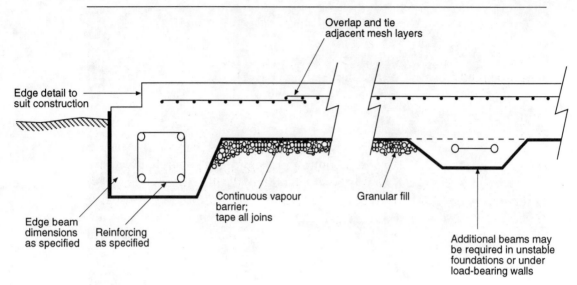

Figure 5.24 *Typical concrete slab construction includes reinforcement and shaped forms. Details will be specified by the engineer*

can therefore be poured independently of the footings, or integral with them. Sometimes the building frame is completed, with wall cladding installed, prior to pouring of the slab. All that is required is to ensure the finished level of the footings is compatible with the finished level of the slab, both being above the natural ground surface surrounding the building. The thickness of the slab (100 millimetres for general use, up to 150 millimetres for heavy machinery or storage use) needs to be accounted for when determining finished levels.

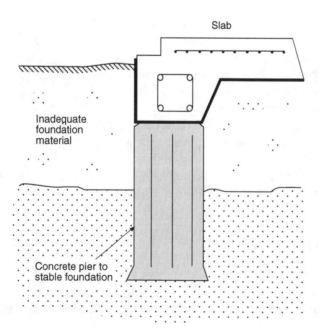

Figure 5.25 *Additional piers may be necessary to found the slab on deeper, more stable material*

In other types of construction, the walls of the building carry load and the slab is designed with a continuous strip footing under the wall. This is either poured as an integral part of it or sometimes separately. Additional reinforcement is required. The same applies to interior partitions which are load bearing. When excavating for strip footings, keep the bottom level, sides vertical, and remove loose material from the excavation to ensure load is adequately transferred to the foundation material.

If wall loads are great, or the ground conditions less than ideal, the strip footing is extending to become a continuous reinforced concrete beam under the loaded area (usually the perimeter edge). If ground conditions are unstable, differential settlement of the slab can be minimised by constructing a number of such beams within the slab, to produce a raft type construction. If necessary, piers or piles can be installed to more stable foundation material at intervals along the beams.

Raised floors

A raised floor may be required for a number of applications such as livestock housing, loading platforms, storage sheds, and shearing sheds. The load conditions acting on the floor need to be considered for each application, since they will be substantially different from standard buildings. These loads are published in various design manuals, so are not too difficult to find. They relate to dead and live loads only for fully enclosed buildings.

The floor material and strength is selected to suit the application. Tongue and groove boards, or sheet timber material is available in a range of thicknesses where a wooden surface is preferred. Slatted timber is common for sheep housing. Woven and expanded metal mesh is available for livestock housing, stairways and walkways. Special materials are also available, for instance concrete slatted flooring for piggeries.

Flooring needs to be sufficiently strong and supported at close enough intervals to remain stable under load, without deflection or bouncing. For low duty flooring, 19 millimetre hardwood tongue and groove boards, or 19 millimetre particle board sheets (which are also joined by a tongue and groove) would need to be supported by joists at 450 millimetre spacings. Plywood is sometimes used.

Standard low duty construction would require these joists to be 100 x 50 millimetres hardwood, themselves supported by bearers at 1.5 metre spacings. Heavier load conditions would require thicker (and therefore stronger) flooring materials and stronger joists supported at closer intervals. Alternative structural supports, such as webbed steel joists, and combination timber and steel joists, are available where the joists and bearers need to have longer spans. This reduces the number of supports.

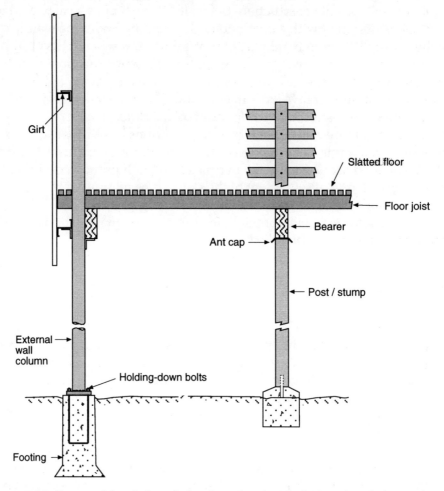

Figure 5.26 *One possible sub-floor design for a shearing shed. Note that the bases of columns are protected against dung accumulation*

Long span joists may need stiffening by using a solid bridge or herringbone struts. Take care when installing floor supports that the floor will finish level and flat, as well as being free from gaps. In slatted flooring the size of the gap must be constant. Most importantly, consider the expansion and contraction of timber floor boards as their moisture content changes. Allow some time for the timber to season before securing to joists. If moisture content is too high they will shrink and the gaps will open up. If too low, the boards may swell later, causing them to bow upwards.

Floor loads are transferred to the foundation by a network of columns or stumps, themselves normally standing on a concrete pad. The pad size is sufficient to prevent the column sinking into the foundation. The column could be made of timber, steel, reinforced mass concrete, concrete block or brick. Selection of material may be influenced by those used throughout the rest of the construction, convenience, or its ability to withstand

Types of construction

Figure 5.27 *Under the floor of this shed, heavy timber bearers have enabled long spans to be achieved, reducing interference to pens under the floor. Note the use of steel posts where height requires extra strength and stiffness. A simple bridge between bearers adds to their stiffness as well*

deterioration. Size and strength will be determined by the magnitude of the floor load, and the height of the column (ie the height of the floor above ground level). Where the height is large enough, long span joists and bearers become cost effective by reducing the number of columns needed.

Figure 5.28 *Under the floor of this older shearing shed the timber stumps show some signs of deterioration at ground level. Note the use of packing to maintain the level and the bearers joined above points of support*

CHAPTER 6

Cladding and building protection

Nearly all farm buildings use corrugated steel sheeting for roof and wall cladding. Standard profile corrugated steel is common but increasing use is being made of alternative ribbed profiles, either for appearance, greater strength, or special applications. Cladding is protected by galvanising (a zinc coating), Zincalume® coating (zinc and aluminium), or Colourbond® coating. This chapter provides a general description of cladding materials and their installation, as well as windows, doors and other aspects of finishing a building. The product manufacturer should be consulted for specific details.

Cladding

A range of cladding materials is available Their correct installation is an important aspect of farm building construction.

Selection

Standard corrugated steel sheet has 16 mm corrugations, a total sheet thickness of 0.47 mm and covers a width of 762 mm (the sheet is wider, but a one and a half corrugation overlap is required).

When installing new cladding, it would be purchased or supplied in full length sheets, cut to measure as required. On a roof it can span up to 1200 mm between intermediate purlins, but only 900 mm at the ends of the sheet. On walls, the maximum span between intermediate girts is 1900 mm and 1350 mm at the ends.

Galvanic protection against corrosion is provided by a zinc or zinc and aluminium coating, 0.05 mm thick. This layer is easily damaged if the sheet

Cladding and building protection

is mishandled, from walking or dragging objects over the sheet, or from abrasion by the metal particles present after installing the sheet. Careful handling and working practices are required. It is important that lead, copper and plain sheet are not in contact with a Zincalume® sheet, despite these materials being in common use for flashings, and other building hardware. Even water discharging from them may cause staining. Zincalume® is not recommended for highly corrosive environments such as piggeries.

Prepainted Colourbond® material is etched, primed and finished with an oven baked enamel in the factory, in a wide range of colours. Other building products are available in the same range of colours.

A variety of other profiles is available such as Trimdek Hi-Ten®, Spandek Hi-Ten®, and Klip-Lok Hi-Ten®. These products have deeper, rectangular ribs, sometimes separated by a flat tray. Heavier gauge steel is also available. These profiles give greater strength to the cladding, enabling it to span greater distances between purlins and girts. Some people prefer their

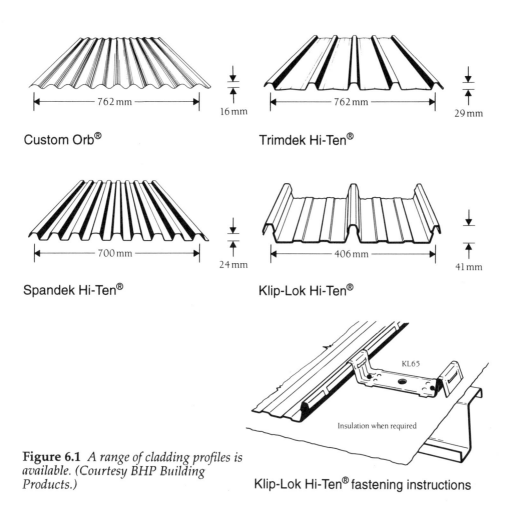

Figure 6.1 *A range of cladding profiles is available. (Courtesy BHP Building Products.)*

Klip-Lok Hi-Ten® fastening instructions

appearance to corrugated steel. Klip-Lok Hi-Ten® cladding is one of a type which is attached with concealed clips. The clip is attached to the purlin or girt, but no screws are used to attach the cladding to the clip. Many of these profiles are required for flat roofs (standard corrugated iron is limited to a 5° slope), but are equally useful in other applications. A different grade of corrugated iron is used when rolled curves are required. Refer to the manufacturer's product information for specific details.

Be aware that second grade cladding is available, at reduced cost, but is likely to have defective protection or be dimensionally inaccurate.

Translucent sheeting, of a number of different materials, is available to provide natural lighting into farm buildings. Special handling and fixing techniques are often required, for instance pre-drilling holes for fasteners. Refer to the manufacturer's specific recommendations.

Handling

Cladding is supplied in packs up to one tonne. Take care to avoid damage when unloading packs, and handling individual sheets. Packs should be adequately supported when on the ground awaiting erection. They should also be kept dry. Moisture collecting between the sheets will quickly cause a characteristic white discolouration, possibly affecting the protective coating.

Figure 6.2 *This cladding shows deterioration, probably caused by moisture collecting between sheets prior to installation*

The edges of the sheets will be sharp. Use dry gloves when handling them. New sheets are likely to have an oily film over them, left over from their manufacture, which might make them more slippery than expected.

Correct work practices are important to prevent damage to the sheeting; walk only over the supporting purlins, spreading weight over more than one corrugation (except for narrow ribbed profiles, where you walk in the trays). Use soft footwear, but avoid rippled soles that can catch abrasive particles. Avoid dragging tools or equipment over sheets and clean up metal drill particles, unused screws and all other debris at least once a day. Cut sheets on the ground wherever possible. (See also the section on Safety on pages 9 and 10.)

Installation

The preferred method of fixing cladding to its supports is with self-drilling, self-tapping screws. These are installed using an electric powered drill, preferably fitted with an adjustable clutch (to avoid over-tightening the screw, damaging the sealing washer, distorting the sheets and overloading the motor) and a reverse setting. A large variety of screws is available to suit steel or timber supports, and in different lengths to suit valley fixing, or the different rib heights for crest fixing. There are also different sealing washers, cyclone washers and so on. Alternative methods include nails, hand screws and hook bolts. Self-drilling screws are quicker, easier and offer a far superior connection compared to nails.

The ends of standard corrugated iron sheets are screwed every second corrugation (five per sheet), the intermediate supports every fourth corrugation (three per sheet). Different profiles, and translucent sheeting, have their own requirements. Refer to manufacturer's specifications.

Operators should be aware of the procedure required before starting cladding. One edge of the sheet has an overlapping edge, the other edge is an underlapping one. A one and a half corrugation overlap is required. If there is a dominant prevailing wind direction for the site, the direction of the overlap needs to be considered (Figure 6.4).

Start the first sheet on the downwind end of the building, with the overlap edge on the outside. Ensure it is square, and with the correct amount of sheet protruding at the gutter end (60 mm with no wall sheeting, 75 mm if the side wall is sheeted, to ensure run-off goes into the gutter). Some slight overhang may be required on the outside edge. Fix this sheet on the outside edge.

The next sheet is positioned with correct overlap, and fixed top and bottom on the overlap, after checking its alignment and squareness. Additional sheets are positioned in a similar way, continuously checking for squareness, particularly as each bay is approaching full cover. The rest of the screws are attached once some of the sheets have been pinned. This procedure is recommended to help stop the sheets spreading during fixing, or spreading

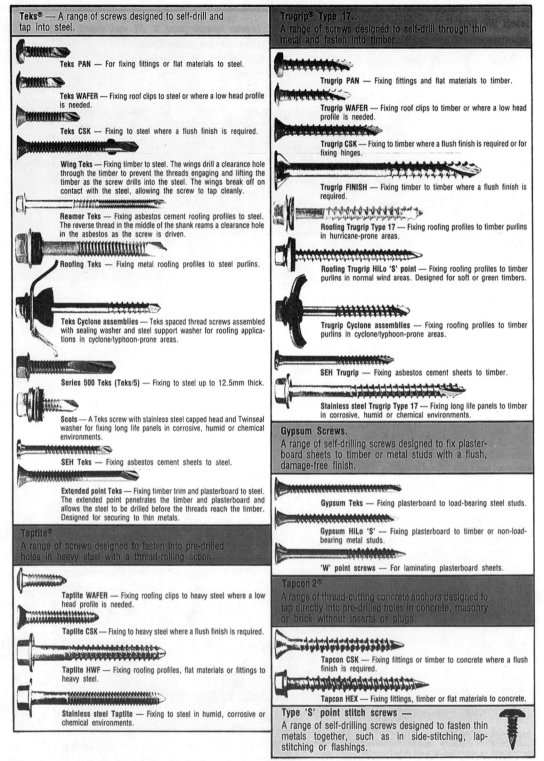

Figure 6.3 *A wide range of self-drilling fasteners is available. This is from one product manufacturer. (Courtesy W A Deutscher Pty Ltd.)*

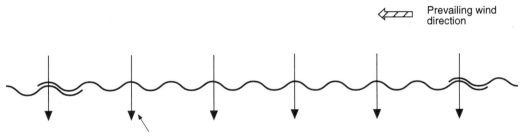

Figure 6.4 *Installation of corrugated iron roof sheeting*

unevenly, and so losing their squareness. However, on large jobs where weather conditions can change part way through the job, it may be preferable to complete the operation in stages (provided you are happy that the last sheet will still be square, and not leave a gap because of insufficient cover). Mark the line of the screws by using a string line or straight edge.

Condensation

This is a problem in many farm buildings, resulting in dripping of moisture from the roof into the building, particularly with flatter roof profiles. To avoid this, it is necessary to prevent moist air contacting the underside of the cladding, and so preventing condensation from forming. Rarely would

Figure 6.5 *Construction in progress and cladding near completion*

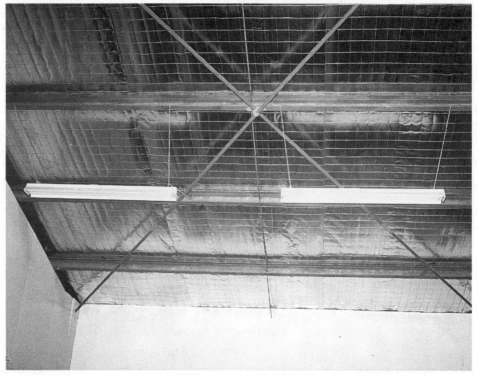

Figure 6.6 *This is a view looking up at the ceiling of a structure. Reflective foil supported by a wire mesh has been installed under the cladding. Note also the cross-bracing of the frame*

farm buildings have a ceiling, but one option is to install reflective foil under the cladding. To be fully effective, it should be installed loosely, to provide an insulating air gap between foil and cladding, with joins in the foil lapped and taped.

Insulation

The reflective foil mentioned above will also assist reduce the heat load in summer for fully enclosed building, and is worth considering in applications such as livestock housing or shearing sheds. Bulk insulation (fibreglassbatts, polyurethane or polystyrene) provides better coverage in winter, as well as providing some summer protection. Polystyrene filled construction panels are available for structural applications. Polyurethane foam is sprayed onto the inside surface of the building after completion, whereas fibreglass batts need to be installed as the cladding goes down. A sandwich of fibreglass and reflective foil is suitable for many applications, supported under the roof sheeting by wire mesh.

More detailed investigation of insulation is required for special applications such as intensive livestock housing. Some of these are discussed in Chapter 9.

Cladding and building protection

Figure 6.7 *Typical window installation. Note the use of flashings to assist weatherproofing and to provide a neat finish to the cut edge of the cladding*

Doors and windows

Most farm buildings now use pre-fabricated aluminium frame sliding windows. These are maintenance free (apart from glass breakage), low cost and can be easily installed in virtually any position required. Installation involves simply screwing the window frame into position. Smaller windows installed against square ribbed cladding can be attached to the cladding, but larger windows need to be supported by a girt, which dictates their height above floor level. The top and bottom of the window are weatherproofed by head and sill flashings respectively, and the sides by a liberal application of silicone sealant. The window frame is pre-fitted with rubber sealing strips. Where the interior of the window is to be finished for a neat appearance, jambs, reveals and architraves are added.

Louvre windows are also in common use. The window frames are simply fixed to jambs fitted into the building frame, and weatherproofed by suitable head and sill flashings.

Personal access (PA) doors are easily fitted at the location of choice. The door jamb is installed in the frame as required, checking for square. The door frame is asembled, sheeted, and fitted with hinges and locks, then fixed to the jambs.

Sliding doors provide for a wide and high opening for access by machinery and produce. The door is suspended on adjustable carriages running in a track. The track is attached to the frame by brackets, and a guide is located at floor level.

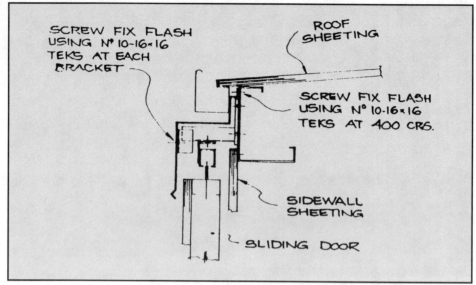

Figure 6.8 *Typical installation detail, in this instance a sliding door, from the manufacturer's handbook. (Courtesy Tamarang Engineering Pty Ltd.)*

Roller doors are carried on brackets attached to the building frame, with guides attached to the jamb to locate the edges of the door. Roller and sliding doors are difficult to make dust-, vermin- and draught-proof. If these conditions are important it may be necessary to fit rubber sealing strips to the perimeter of the door opening.

Weatherproofing and roof drainage

Ridge capping is screwed or riveted to the roof sheeting at every second corrugation. The valleys or trays of the roof sheeting are manually turned up at the ridge end to help prevent rain being blown up under the ridge capping.

A barge capping is screwed on at the junction between roof and end wall, positioned in conjunction with the corrugations or ridges to prevent rain being blown under.

Where the barge capping meets the ridge capping, a pre-fabricated mould could be used to keep weather and birds out. Alternatively, the capping can be trimmed by hand to overlap, then sealed with silicone and screwed down. The corners of adjoining walls are also fitted with corner capping.

Gutters and downpipes are recommended on all farm buildings, to assist with site drainage and to collect rainwater. Various styles of guttering are available, and they are fixed with clips or screws as required. Downpipes can deliver water to storage tanks, but if not, ensure downpipe discharge is drained away from the building. Secure downpipes to the shed wall with straps.

Cladding and building protection

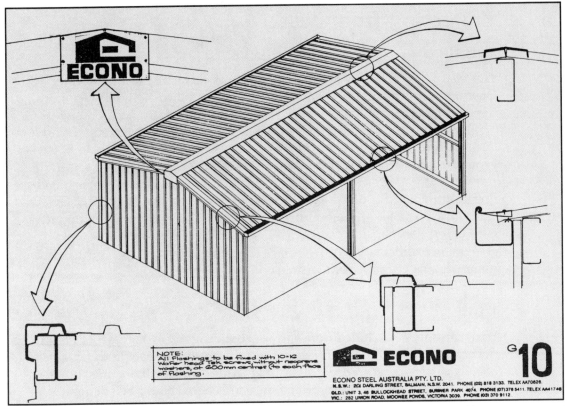

Figure 6.9 *Typical application of weatherproof capping. (Courtesy Econo Supershed.)*

Figure 6.10 *Although spilling to the ground, this roof drain is well installed, and delivers run-off away from the building site via a graded concrete drain*

Protection from pests

Termites

The treatment of timber for termite attack has already been mentioned in Chapter 2. Soil treatment is an option prior to construction, but is not preferred for many rural applications. Follow-up soil treatment is possible.

Termites can be quite persistent, and can travel undetected from the ground, through concrete or brickwork parts of a structure into an above-ground timber frame. Ant capping of piers and sub-walls is an essential protection measure. However the ant cap does not prevent termites travelling around the edge of the cap; it requires them to be more readily detected during inspections of the building. Hence, regular checking for the presence of termites is still required.

Where termites are a problem, it may be preferred to use construction materials that are safe from termite damage.

Birds

Birds can only be excluded from enclosed buildings by installing bird netting over all possible points of entry, particularly around the roof line and eaves, or fully closing those parts with capping. Windows and doors must be close fitted, but normal weatherproofing measures here should be adequate.

Figure 6.11 *Typical use of bird netting*

Cladding and building protection

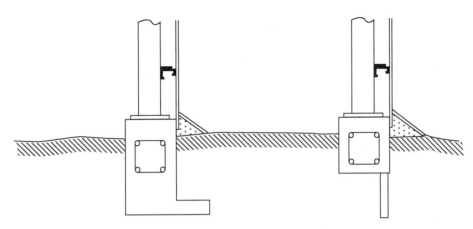

Figure 6.12 *Concrete footings incorporating "rat wall" extensions.*

Rodents

As a first step to avoiding rodent infestation, remove harbour that might provide nesting sites from around the building. Footing designs can be modified to help prevent rats burrowing under buildings by incorporating a horizontal or vertical "rat wall" in their design. Small gaps anywhere

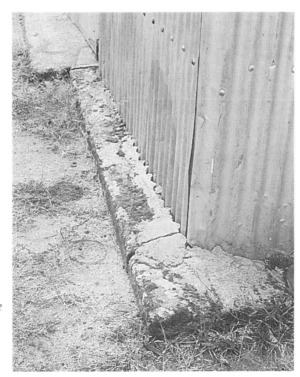

Figure 6.13 *Concrete at the base of this seed shed wall is intended to keep mice out, but note they can get through the large cracks in the concrete*

should be filled in. Drainage outlets can be fitted with simple check flaps to prevent rodents gaining entry.

Walls should be finished neatly, with capping applied correctly. Concrete aprons around the external perimeter of the wall cladding may help prevent burrowing. Insulation materials also provide ideal nesting sites, and so should be sealed or connected correctly to the cladding.

Windows and doors need to fit securely, and the gap around and below doors will need to be blocked by modifying the jamb and providing a kick plate.

Locate girts with the open side down to minimise nesting sites, and fill any hollow parts of the structure. Cavity walls should be protected by wire netting at their base to prevent rodents climbing up the cavity into the roof. Enclose electrical wiring in conduit to protect it against chewing.

CHAPTER 7

Building services

The installation of services to a building (electricity, drinking water from town supply, waste water drainage) is performed by licensed tradespersons. This chapter discusses the general requirements of these services, and procedures required for their connection.

Electricity

All electrical work must be performed by a licensed contractor. They can also give advice at the planning stage in regard to the type of service and outlets required, and expected cost. The local electricity supply authority can provide advice in regard to the supply of electricity and transformer, if none is available in close proximity, and for situations where the supply must be upgraded to cater for an increased load. In some cases, siting the building nearer the electricity supply could be an option.

Predict the maximum load for each of the outlets, and the requirements of any possible expansion, at the planning stage. There is a limit to the amount of current permitted along power lines, so there is therefore a limit to the capacity at each outlet, and of the total supply. Additional load which exceeds these limits will require an expensive upgrade to the supply.

Large electricity consumers may be able to take advantage of attractive tariffs offered by supply authorities. Consider this aspect at the planning stage, since it may influence the nature of the equipment in use, or the hours of its operation.

Consider the location of lights and power outlets in the building. It is far easier to put in a few more at the construction stage than go back later on. Installing inexpensive flood or spot lighting outside the building is a major advantage for loading areas (including stock) and for safety and security.

Figure 7.1 *Simple exterior flood lighting will greatly assist night work. These lights are fitted with a movement sensor. The container outside the shed houses a cool room for this small vineyard*

Good internal lighting is essential in work areas, and in specialised structures such as those for livestock housing. Various types of lights are available; the choice may depend on a variety of factors including cost, running cost and heat output, as well as light intensity, wavelength and evenness of distribution.

Electricity will also be required to run ventilation fans, workshop equipment and so on. It is preferred for refrigeration and water pumping, and is an option for water heating and general heating.

Engine driven generators are used as permanent installations in remote areas, as back-up units in the event of power supply failure for critical activities, and as portable units where permanent supply is not required. Where a building is to be connected to electricity, a temporary connection can often be arranged to provide power to construction workers.

Batteries of 12 or 24 volt d.c. are an option for remote low duty applications. Such systems can easily be coupled to solar, wind or turbine powered battery charging systems, although the cost per kilowatt is reasonably high.

Water

Stormwater drainage

The importance of good site drainage has already been discussed (see Chapter 1), as has the installation of gutters and downpipes (see Chapter 6).

If the roof run-off is collected in tanks, note that for each millimetre of rain, one litre of water is possible for each square metre of floor area (not roof area), although the first few millimetres of rain never seem to make the tank due to initial wetting of the roof surface, imperfections in the fall of the gutters to the downpipes, and evaporation. The size of tank you put on the shed needs to be matched to the size of the roof, the rainfall for the district and the expected consumption rate. As a rough guide, work on the basis of 180 litres per person per day for domestic use with a septic tank, such as would be required for an amenities building including toilets and showers.

Water supply

A licensed plumber is required to work on town water supply systems. Although not essential for applications where the water supply is independant from town supply, their advice is preferred to ensure the correct supply volume and pressure to outlets, and to avoid problems with plumbing in the future. For most applications, the use of plastic pipes is widespread, making do-it-yourself plumbing somewhat easier, provided the limits of these materials are not exceeded.

Where multiple outlets are required, such as for livestock housing, it is essential that peak flow rates be considered in detail, to ensure that the volume of water delivered to each outlet is sustained during critical stress periods.

Water may also be used for evaporative cooling of livestock, produce and plants; for irrigation in greenhouses; for washdown water in dairies and processing areas; and for the fast filling of spray tanks. Specialist advice is suggested to predict annual volume requirements, peak flow rates, pressure distribution, and installation details.

Wastewater drainage

Some rural situations call for the use of septic tanks and absorption pits for containment and partial treatment of wastewater. An application listing the capacity, location and design details of these must be submitted to the Health Department through the local council. Full details are available through those channels. Farms on the outskirts of larger towns may have sewage connection available and pumping wastewater into the sewage system is then an option. In most rural situations, little effort is made in regard to wastewater disposal, but this is becoming increasingly unacceptable.

The disposal of effluent from a designated development (such as intensive livestock housing) is a specialised topic, outside the scope of this book. Information on effluent requirements under these circumstances should be attained from the relevent planning and regulatory authorities.

Access

The importance of good vehicle access to most farm buildings has been discussed in Chapter 1. If the building or development is likely to attract passing traffic (such as a roadside stall, weighbridge, pick-your-own operation or even traffic associated with large numbers of staff), council is likely to require some provision for getting vehicles on and off the property safely. For minor roads, this may be as simple as offsetting the entrance gate from the line of the carriageway. It may be much more complicated and expensive on main roads. Again information on these requirements must be finalised with local council.

Ventilation

For fully enclosed farm buildings, internal temperatures in summer can become extreme, even with insulation installed. Ventilation is used to remove heated air and replace it with air from outside. Natural ventilation relies on air movement only by convection, whereas forced ventilation uses fans to move air through and out of the building.

Natural cross-flow ventilation is the easiest to provide, usually from side-to-side through open windows or doors. This method is suitable for general purpose buildings, which are not too wide (less than ten metres) and which are not too sheltered by terrain, trees or other buildings. Where cooling requirements are more significant, such as in greenhouses and livestock housing, the building is constructed with more substantial ventilation methods:

- Side walls can be constructed with much of the wall area able to be opened for ventilation.
- Fans to push air into the building, sometimes using ducts to distribute the air more uniformly.
- Fans, including non-motorised "whirlybirds" on the roof, to suck air out of the building.
- Ridge vents to facilitate natural convection during calm weather, with standard saw-tooth type of frame construction. A roof pitch of 15° is preferred.
- Water to provide evaporative cooling of the air prior to its entry into the building, or by applying water directly to the animals or plants.

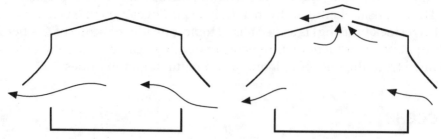

Figure 7.2 *Natural ventilation can be improved by incorporating a ridge vent, particularly in wider buildings, or where still air conditions prevail*

Building services

Figure 7.3 *This poultry shed has wall mounted blinds and a ridge vent to control ventilation*

Figure 7.4 *A ridge mounted polythene duct distributes air through this greenhouse*

Figure 7.5 *"Whirlybirds" can assist ventilation, without requiring major alteration to the roof structure*

CHAPTER 8

Inspection and maintenance

Initial selection of materials, and good siting, design and construction play a large part in minimising the amount of maintenance and repair needed over the life of a building.

A good location provides good drainage, stable foundations and perhaps some protection against wind loads. All these are important in preventing movement of the building and subsequent problems. Materials should be selected with strength and durability in mind, not just the purchase price. Structural design of the building should ensure its integrity for decades. Sloppy work practices during construction, perhaps for the sake of a few hours extra labour, can lead to continuing maintenance and repair problems.

However, there will always be some routine maintenance and repair tasks required for farm buildings, whether they are newly constructed or still in service long after their original design life.

Inspections

Include maintenance inspections into the work schedule, in the same way that plant and equipment maintenance is done, and fix small problems before they become big problems. Keep a supply of materials and hardware on hand, so that maintenance and repair jobs can be done easily and quickly.

Check visually for deterioration of building components on a regular basis. In particular, look for corrosion of steel and deterioration of timber that will affect the structural integrity of the frame. In many cases, a visual inspection is not sufficient; for instance corrosion can be much worse than it looks from the outside, concealed by paint, or appearing to be just surface flaking when severe pitting is underneath. Termite damage is not visible

Figure 8.1 *This shearing shed needs replacement glass instead of bags to keep the rain off the sheep*

from the outside surface of timber. Do not be too afraid of taking a screwdriver or knife to suspect areas, testing their appearance. It is better to know there is a small problem before it gets worse. Tapping with a light hammer can often reveal internal problems.

If the building is associated with a corrosive environment such as fertiliser storage, a shearing shed sub-floor, piggeries and so on, a rigorous inspection regime is essential.

Figure 8.2 *These damaged floorboards should be repaired promptly*

Figure 8.3 *This guttering has been damaged, allowing ponding near the footings. The loose edge of the roof iron should be screwed down before more damage occurs*

Maintenence and protective measures

Corrosion

Check for corrosion and deterioration at structural points of connection. Rust in bolts and welds is common, and even if the main components are sound, failure can still occur. The end grain of timber is notorious for localised deterioration.

Deterioration will be accelerated in the presence of water. Look for areas where drainage is poor, allowing water to pond close to footings, support columns, exposed timber, and under the building. Correct such drainage problems as they occur. They could be caused by insufficient grade away from the building, or defective gutters and downpipes: check the performance of these occasionally when it is raining, and keep them free of debris. Extra concrete, tapered outwards on the top, can be added to concrete footings to keep water away.

Steelwork should preferably be galvanised. Where the galvanising is damaged by welding, cutting and drilling, or by accidental damage, it must be painted with two coats of zinc chromate paint, which provides much greater protection than standard primer. Be careful to avoid using materials which might be incompatible; for instance lead can react with some zinc and aluminium products.

If black steel is used, paint it with one coat of good quality primer and two coats of paint for maximum life. This will need touching up occasionally, especially on the critical parts of the structure where any corrosion must be avoided. In corrosive environments, this will require a far more rigorous approach, ensuring adequate selection and application of appropriate coatings.

Walls and cladding

Secure any loose cladding against the wind. In older buildings, where the cladding is nailed onto timber supports, the nails will come loose as the timber shrinks and swells. An occasional loose nail may not appear very important, but during high wind, flapping of part of a sheet can loosen adjoining nails, and the whole sheet is then at risk.

Replace loose nails with self-tapping screws. Another nail would be a waste of time. Trying to attach cladding to a rotten batten or purlin is also of little use; a replacement is required. One possibility is to locate the new support immediately next to the old one, fixing it to the main frame with a strap or tie. The loose sheet, and the adjoining sheets, can then be screwed down quite easily.

A minimum of around 600 millimetres clearance is required (more if the building height exceeds 2.4 metres) between the building wall and any other obstruction to facilitate repairs. Keep this in mind during planning.

Structural supports

Damaged structural members should be repaired or replaced immediately. Their failure during strong wind, which could occur at any time, may cause

Figure 8.4 *The post has been bumped out of plumb, and the bolted connection between truss and purlin has separated. This has been partly caused by end grain deterioration of the purlin*

collapse of the whole structure. If time does not permit major repairs, apply temporary bracing, tie downs, struts and so on as required.

For portal frame constructions, occasionally check that the cross-bracing is reasonably tight, and adjust it as required. If left loose because of stretching, expansion or poor installation, the building frame has the potential to move, weakening structural connection points or failing under stress.

Subsidence

Do not let a building lean or sink too far before attempting repair. Subsidence is not usually a failure of structural components, but of the foundation. If allowed to sink too far, great stress can be placed on the frame and its points of connection, which might lead to structural failure.

Leaning can sometimes be corrected by first straightening up the building using a chain block or jacking then applying some bracing. If the lean has been caused by failure of some original bracing, then it will need to be re-braced with a superior material. This assumes that the foundation allows the frame to move back up again, and you do not bend or break the frame. Pulling on the frame with a tractor and chain lacks the sensitivity required to check if the procedure is working. Also, the line of pull is fairly steep, which risks upsetting the tow vehicle. A good ground anchor is preferred.

Sinking can usually be corrected by jacking the column/s affected, then treating their foundations and/or bases. This may require excavation around the column, which is possible if the frame is jacked and braced securely. If the foundation is soft, a wide plate can be installed under a stump or pole, or more concrete added to a more substantial width and thickness over an existing concrete pad.

If the sinking was caused by deterioration of the base of the column, it could be repaired, provided the repair possesses the same strength as the original column, in all directions. If the structure was a pole-in-ground type, cutting off the bottom of the pole removes its ability to withstand movement. If a concrete footing is made to take its place, the whole frame will need to have cross-bracing installed, probably in two directions.

In all these cases, investigate the cause of the problem; there is a fair chance it is related to poor drainage, so unless the drainage is corrected, your repair will remain temporary. If major structural damage is noticed, expert advice may be necessary.

Timber

In-ground timber should be protected as described in Chapter 2.

Above ground structural timbers are not often painted or sealed in farm buildings, with the exceptions of where the end grain is exposed, and where softwood sawn timber is used. However, painting or some other form of sealing may be necessary to minimise the transfer of moisture into and out of the timber, thereby providing some control over shrinking, swelling and subsequent cracking. To be effective, the coating must be relatively continuous, and the moisture content at the time of painting needs to be suitable (ie seasoned timber). Primer, undercoat and two topcoats are preferred, although some all-purpose acrylic exterior paints can be applied directly. Thorough preparation is required, and reapplication as necessary.

Keep the end grain of timber protected from the weather. A typical example is where timber purlins are used on a building without end walls. A barge capping could still be used to protect the purlins. Do not rely on the end grain to hold fasteners.

When using bolted connections in timber, make the hole slightly oversized, to assist in the location of the bolt, and its removal if maintenance is required. Moisture in timber will corrode bolts quickly, so apply a good coating of grease before inserting the bolt.

Because of timber shrinkage, all bolted connections will need to be checked for tightness one or two months after construction, at the end of the first year, and periodically thereafter. Check nailed connections similarly, to make sure that the timber has not shrunk away from the nails.

CHAPTER 9

Special purpose buildings

This chapter expands on the design requirements of some special purpose buildings, and should be read in conjunction with the previous chapters dealing with materials, structural design principles, and provision of services. Additional specialist advice is recommended prior to construction of new buildings, or major modifications to existing ones. When considering construction of a special purpose building, get out and inspect a range of buildings already in use, and discuss their good and bad features with their owners and operators.

Note that many of the principles behind the structural design of buildings apply to other types of structures such as fences, trellises and artificial wind shelters.

Greenhouses

A wide range of construction methods and covering materials are available for greenhouses. Most also require consideration of the internal environment, which naturally must facilitate plant growth.

Structure

Of all farm buildings, greenhouses probably come in the greatest variety of shapes and sizes, covering all the types of construction described in Chapter 4. Different types of structure vary greatly in cost and their ability to control the internal environment. Choice of structure will also be influenced by the duration of the crop growing season, and whether the crop to be grown is in the ground or on benches. For example, a double skin igloo type construction is relatively inexpensive to construct, but has difficulty in providing adequate natural ventilation to control high temperatures. Such a structure may, however, be quite suitable if its principal purpose is to extend the winter growth period of plants.

Greenhouse environment is relatively corrosive. Galvanised steel is the preferred structural material, although aluminium and timber structures are also in use. Steel is relatively cheap and durable if fabrication defects are avoided and welds and bolted connections are sufficiently protected against corrosion. Steel's smaller cross-section compared to timber casts minimum shadows.

Despite appearing to be light duty structures, the wind loads on greenhouses are as high as any other structure of the same shape and size, so the design of footings is just as important.

Coverings

Glass is the traditional covering for greenhouses, and it still retains the best optical performance (that is, the percentage of light passing through the material, and its wavelength spectrum). Its disadvantages are that it is the most expensive covering, it admits too much energy during summer and loses heat during winter. Structurally it requires a stronger, straight sided frame to support its weight and small sheet size, which casts distinct shadows. Glass is easily damaged by hail.

Consequently, a very wide range of alternative plastic coverings is available, to overcome some or all of these disadvantages. The type of plastic varies, giving different optical properties, some specifically selected for

Figure 9.1 *Traditional glasshouse construction*

Figure 9.2 *Corrugated acrylic sheeting has been used on this greenhouse. Note the use of ridge vents and provision for ducted fan forced ventilation*

certain preferred wavelengths. Most are opaque, which diffuses the light to reduce shadowing. Many are double skinned, to reduce heat loss during winter. They may come in semi-rigid sheet form or as flexible films. There are large differences in cost and lifespan.

Choice of covering material becomes a decision involving many variables, but it comes down to the cladding which meets the required standard for optical performance which also gives the lowest annual cost for the floor

Figure 9.3 *Exhaust fans in the end walls provide ventilation in these plastic film tunnel greenhouses*

area covered (including maintenance and replacement costs of the covering, and the annual supplementary heating costs of the greenhouse).

Ventilation

A major concern with greenhouses is that they get too hot during summer. Incoming radiation can be reduced by shading or by using a partially reflective covering. Even so, hot air contained in the greenhouse needs to be removed by ventilation.

Good natural ventilation is often preferable to forced ventilation, mainly because of the lower cost and lesser dependence on powered machinery. However, to be effective, sufficient roof and/or wall area must be able to be opened; up to 30 per cent of roof area may be necessary. Cross-flow ventilation is not as effective as that from combined side and ridge vents, so the design of the greenhouse plays an important part here.

The ventilation rate may need to be as high as 60 air changes per hour; that is the volume of air contained in the greenhouse needs to be removed up to once a minute. Fan forced ventilation may be necessary if natural ventilation is not sufficient, or if the design of the structure will not permit it. This also allows the option of evaporative cooling to be used as well. Fans deliver air usually end to end, either through low cost plastic film ducting, or by exhaust fans mounted in the end wall.

To maintain an even temperature it is preferred to have vents that are fully adjustable, and their opening and closing automated in response to air temperature. Some quite sophisticated controllers are available to manage this and many other functions.

Figure 9.4 *In this greenhouse a sophisticated drive system automatically adjusts vent opening in response to air temperature.*

Figure 9.5 *Polythene duct ventilation systems can be adapted to include heaters and/or evaporative cooling units*

Some ventilation is required during winter (from two to four air changes per hour) to maintain adequate atmosphere in the greenhouse.

Evaporative cooling

Air temperature can be reduced as it passes into the greenhouse by first passing through a continuously wet pad. The maximum cooling achieved is to the wet bulb temperature of the air at the time, which depends on its relative humidity. A high air flow rate may not achieve the maximum cooling effect because it may not have sufficient contact time with the cooling pads, so a large pad surface area is preferred. Commercial cooling systems are available, which can be combined with plastic film ducting, to distribute cooled air throughout the structure. Alternatively, purpose built pads can be installed at the end of the greenhouse opposite the exhaust fans.

Supplementary heating

Maintaining a minimum winter night time temperature is an important aspect of many greenhouse operations, justifying the use of supplementary heating. Circulating heated water through pipes within the greenhouse is a popular method, enabling direct heating of the plant surfaces by radiation from the pipe, as well as heating by conduction. Alternatively, heated air can be delivered into the greenhouse.

Heating requirements, and costs, are greatly influenced by the amount of heat lost through the covering of the structure. Some covering materials are far superior than others in this regard in that they allow lower conductivity. Double skin coverings achieve this by maintaining an air gap between layers of the covering, either by virtue of the structure of the covering, or by use of a small fan to blow air between the two layers of film. Condensation on the covering also has an effect.

Some greenhouses are fitted with special moveable blankets or coverings to contain heated air to a more confined area surrounding the plants.

Heating requirements can be reduced by taking advantage of day time radiation. Large mass structures made of concrete, rock fill or water vessels absorb heat for re-radiation during night time. Passive solar heating of water, for circulation during the night, is an example.

Fuel for supplementary heating could be electricity, gas, or fuel oil (including waste oil). Choice will depend largely on relative annual costs of fuel and heating equipment. Provision within the structure will be needed for the equipment, its effects and any safety requirements that go with it.

Other environmental factors

Humidity levels may need to be controlled, by ventilation and/or heating to reduce humidity levels, or by misting or fogging to increase them.

Propagation greenhouses have special requirements for environmental control, incorporating special purpose heating methods within the growing medium. Hydroponic crops may also have special requirements. Irrigation within the greenhouse can be accomplished in a variety of ways; spray, mist, continuous trickle tubes, drip and sub-surface methods, and hydroponics.

The floor of the greenhouse may need special consideration for bench and container grown plants; blue metal and concrete (or a combination) are the common options. Cost, disease control and trafficability by picking trolleys, for example, are the major factors.

Access within the greenhouse for picking, spraying and other operations needs to be considered. Some operations use trolleys, conveyors, tramlines or moveable benches to facilitate operations.

Some growers install artificial lights to extend "daylight" hours in the greenhouse. These will need adequate structural strength and an electricity supply. The type of light is matched to the wavelength and intensity preferred by the particular crop

Shearing sheds

General planning

Most of the factors affecting site selection for the shearing shed have already been summarised earlier. In addition, the shearing shed must be located near the sheepyards, and the flow of sheep into and out of the shed,

Figure 9.6 *Dual purpose shed; raised shearing board and pens are features of this shed*

through the yards, must be considered. Where no yards currently exist, the best site can be determined easily, based on stock movement, drainage, exposure and so on. Where a new shed is to be built near existing yards, a lot of thought must go into its orientation and precise location, because the existing yards may not be ideal.

Wool shed size relates to both the physical size of the building, which determines the number of sheep that can be held and the space available to store wool, but also the number of shearing stands, which determines how quickly the mob is shorn.

The best size depends on the maximum number of sheep to be shorn, the number of days in which you would like to have shearing completed, and the likelihood of wet weather during that time. As a rule of thumb, estimate 120 (merino) sheep per shearer, enough woolly sheep for a day inside the shed, and up to another day's shearing housed under the shed or elsewhere out of the rain. Other factors include the capital cost of larger sheds, the optimum number of shedhands to make an efficient team, and whether the shed is to be used for any other purpose.

Because of the cost of a shearing shed, and the fact it is only used a few times a year, there is a lot of interest in dual purpose sheds. Part of the area is taken up with raised pens and board, with the wool room concreted at ground level. The wool room can then be used for storage when the shed is not in use for shearing and crutching, although it will need a good clean before shearing. Some sheds are designed with a working race inside, and so double as undercover working yards should they ever be needed.

Special purpose buildings

Sheds requiring clip preparation accreditation must meet certain standards to eliminate contamination. Construction and finishing details will need to be considered with this in mind.

The environment under a shearing shed is quite corrosive. Careful selection of construction materials is necessary and special precautions are needed where the below floor structure is exposed to moisture.

Sheep access

Entry to the shed is normally from the forcing area of the sheepyards. If these are not adjoining, a simple laneway will do. The ramp up to the shed should be wide enough to allow sheep to move up together. It should have a non-slip surface and solid panels on the sides. It may be stepped if the gradient requires it. Prevent baulking at the top of the ramp by ensuring darkness from under the floor (have the slats across the direction of movement of the sheep), but good lighting above the floor and in the direction the sheep are to move.

Provide for sufficient holding space in the counting out pens, and for the return of shorn sheep back to the working parts of the yards without getting in the way of unshorn sheep.

It is necessary to be able to store woolly sheep under an elevated wool room floor, or an adjoining covered area. Consider how best to get the sheep into, and back from those areas into the shed pens with minimum fuss. Use laneways and gates that are wide enough, and consider the natural behaviour of the sheep when designing these areas. Some sheds, mainly in areas

Figure 9.7 *This shearing shed has a good wide sheep entrance from the yards, with a dividing panel in the middle and blind panels either side. Counting out pens return sheep to the yards after shearing. Good ventilation is achieved through the hinged shutters and open sliding door*

where wet weather is likely during shearing, have internal ramps for moving sheep from below the floor to the catching pens.

Pens and gates

Holding pen capacity should not exceed 100 woolly sheep, to reduce the risk of smothering. Larger storage pens can feed into smaller ones, with the forcing pen and the catching pen holding 20–25 woolly sheep. For most shearers, this will only need filling once during a run, reducing the time spent penning up. Allow 2.7 sheep per square metre.

Good lighting and ventilation are required for the sheep storage area.

A single catching pen per shearer is preferred. It should be no more than three metres deep, to reduce the distance sheep have to be dragged to the board.

Filling the catching pen is often difficult, particularly when trying to force sheep toward the noise at the shearer's door from a forcing pen located behind the catching pen. Alternative penning arrangements are in use, using front fill and race fill layouts so that sheep are filling the catching pen by running away from the board. These designs need to be carefully planned, as they will influence the dimensions of the pens and between the pens and stands, and affect the overall layout of the shed. A few sheds have a conveyor system that delivers sheep to the shearer.

Baulking can be discouraged by a high wall between the catching pens and the board, and by careful arrangement of the floor slats across the direction of sheep movement. Note however that some shearers prefer the slats to be parallel to the direction they are dragging the sheep in the catching pen, to minimise the effort required.

Gate design can greatly influence the ease of sheep movement between pens. Conventional hinged gates need to be swung into the sheep which is difficult when the pen is crowded. Lift-swing, tilt-swing, slide-swing and parallel lift gates can all be used to minimise such a problem.

The board

There are specific requirements for the distance between stands, and the location of the downtube of the shearing machine, in relation to the door to the catching pen and the location of the let-go chute or door. A wider distance between stands is preferred, and it is necessary when there is a catching pen for each shearer anyhow.

A raised shearing board (about 800 millimetres high) can improve the productivity of shed hands, by reducing the effort required to pick up the fleece and clean the board (a batten is used rather than a broom). A raised board can increase construction difficulty and therefore cost, but is an integral part of dual purpose sheds, and worth considering.

If there are three or more stands, another option is to build a curved

Special purpose buildings

Figure 9.8 *Lift-swing gates are one type that help shed hands pen up. This shed has translucent sheeting in the south end wall only, to give good natural light but minimise direct sun. Note the high wall between the catching pens and the board. Light shining up from under the floor slats can baulk sheep, but the slats are arranged to be across the direction of sheep movement. This shed also utilises an alleyway to get sheep to the catching pens quickly*

board. This also has greater construction difficulty, but can make front fill catching pens easier to incorporate in the design, and reduces the distance shed hands must walk in delivering the fleece to the table. Having two tables is another option. (Figure 2.8 on page 43 shows a curved board.)

The surface of the board should be sound, to give a safe and solid working surface that is easy to sweep and clean. Some operators coat the timber, but be careful not to make it slippery.

There are two main methods of releasing the sheep. In across-the-board sheds, sheep are let out through an opening on the outside wall of the shed, on the opposite side of the board to the catching pen. This has the potential to interfere with other activities on the board, and has been superseded in new shed designs by a let-go chute beside the shearer (centre-board layout). The chute leads to a counting out pen below the shed which is connected to the outside yards. A variation for wet weather situations is to build a return race in the shed, but this detracts from the storage capacity of the shed.

Good lighting and ventilation are required on the board. Natural lighting through skylights are likely to make it too hot. Try and avoid draughts blowing up the chute. Provide suitable places for the storage of the shearers' tools and gear, preferably where it won't fall down onto the board or down the chute.

Figure 9.9 *Older style shearing shed, with timber frame construction. The centre pole interferes with activity during shearing, and the low roof and lack of ventilation make it unpleasantly hot during summer*

The wool room

A well laid out wool room will assist good clip preparation and keep wool away from the board. Dual wool tables, sufficient wool bins in close proximity, and loading the main line directly into the wool press will all help, and so the placement of these should be considered carefully. Good lighting and ventilation are required. Provide a floor surface that is solid and safe to work on, and easy to clean.

Although adding to shed cost, there should be sufficient storage area for the baled clip. Many people cope by stacking bales on top of each other, or loading the bales onto trucks for temporary storage elsewhere. However this requires additional work, often when the shed is busy. An elevated wool room floor will make truck loading easier, otherwise they will need lifting by hoist or front end loader.

Amenities

Workers deserve a pleasant work environment, with suitable amenities. A toilet and washing area is necessary, and a separate lunch room is preferred. All should be equipped with appropriate facilities.

For safety, the grinder should be securely enclosed in an experts room, to protect workers from sparks and wheel shattering. Provide good lighting and ventilation. Where stairs are required, up to the wool room or to a raised board, they must be correctly designed, with hand rails as required.

Grain sheds and silos

Bulk stored grain imparts substantial sideways forces to the walls of the structure containing it. As well as the static loads arising from the mass of the grain, there are additional loads generated when the grain flows during filling and emptying.

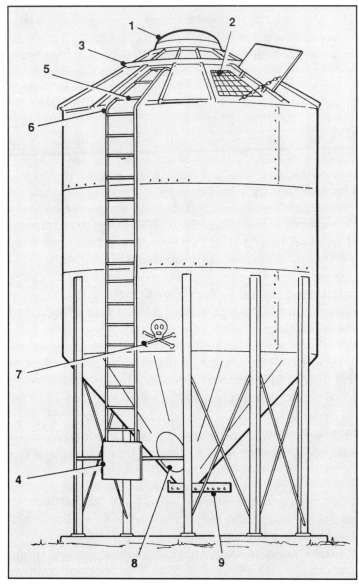

1. Centre inlet guard
2. Roof aperture (where fitted) with a guard with a child-resistant catch
3. Roof safety rail
4. Hinged ladder cover and catch
5. Ladder rungs up roof
6. Ladder stiles curve onto roof
7. Safety signs
8. Base access (manhole)
9. Grain discharge catch (where necessary)

Figure 9.10 *A typical silo showing required safety features. (Source: NSW Workcover Authority Code of Practice; Safety Aspects in the Design, Manufacturer and Installation of On-Farm Silos and Field Bins, 1991.)*

For bulk grain sheds, wall frame components must be substantially stronger than general purpose sheds, with the cladding secured to the inside of the frame. Some sheds require external buttressing to transmit the sideways load to the ground. Some bulk sheds have a moveable roof to allow tipping semi-trailers full access.

Silos utilise the tensile strength created by a circular steel ring structure, and can be thin walled with little additional framing. Yet for this to be stable, the depth of the grain against the silo wall must be uniform; silo failures have occurred with uneven grain depth due to loading and unloading at the edge of the silo rather than the centre. Such a practice also places uneven load on the footings below raised silos, possibly contributing to uneven settlement.

The footings must be designed to transfer the loads from a full silo to the foundation without sinking. An empty silo is also subject to high wind loads, and the footings must accommodate this condition as well. Follow the manufacturer's recommendations regarding footing design, and anchoring.

Flowing grain can suck an unsuspecting person down into it, causing suffocation. Modern silo and field bin design requires mesh guards to be fitted to all openings in the roof of the silo to prevent entry to the space above the stored grain. Note that this space can be hazardous also from dust and chemical fumes, and a silo is defined as a "confined space" under various regulations. These require specific precautions. Access into the silo for cleaning should be provided by a port or door in the base which is inward opening (to prevent accidental opening when the silo has grain in it).

Ladders on the silo must meet particular standards, and a grab rail should be fabricated onto the top of the silo. The bottom of the ladder should be fitted with a hinged or removeable (but child proof) cover.

In operations where multiple silos are in close proximity, carefully plan their location to maximise grain handling efficiency. Provide good access for large trucks and clear height for tipping trucks and augers. Keep well clear of overhead power lines.

Intensive pig housing

Materials

The environment in intensive pig housing is highly corrosive. Materials must be selected for durability. Concrete must be of high strength and quality, with best practices employed for all aspects of the job. Specify 40 MPa strength, 80 millimetres slump, sulphate resistant concrete, and supervise all work closely. Ensure formwork, reinforcing, placement, jointing, finishing and curing requirements are all adhered to.

Steel components should be galvanised, with strict attention to installation and additional protection details. Sub-walls of brick or concrete block are common.

Shed layout

Each class of pig has particular accommodation requirements; boars, dry sows, farrowing sows and piglets, weaners, pork and bacon growers. These requirements relate to pen size and stocking density, as well as optimum environmental conditions, water and feed requirements. Consult specialist sources for details. Loading areas and laneways for pigs also have particular requirements. Pigs have good peripheral vision, and are easily baulked, so laneways should be narrow, with solid panels on each side. Around 20 pigs is enough in a single force pen.

Multiple shed piggeries should be built with the buildings separated by a reasonable distance. This improves the performance of cross-flow ventilation by minimising sheltering. It also reduces the risk of transmission of air-borne pathogens.

An east-west shed orientation is preferred, and given the problems of summer cooling, wide eaves help provide shade to the inside wall.

Ventilation

The general principles of natural and fan forced ventilation have already been mentioned (see Chapter 7). Ventilation is extremely important for pigs, as summer heat stress is a critical factor in pig performance (and survival, in extreme conditions). Housed pigs themselves generate large amounts of body heat. Air movement around the pigs is therefore an essential aspect of temperature control during summer. The structure must contain large areas of ridge and cross-flow ventilation, which will often limit the width of the shed.

Ventilation must be accurately controlled to match the conditions, and not be draughty in cold weather.

Ventilation is also required to remove excess carbon dioxide, hydrogen sulphide and ammonia gasses which will accumulate in the shed, and to keep humidity less than 80 per cent. Good ventilation also assists with control of air-borne diseases.

Cooling

In high ambient air temperature, air movement about the pig will be insufficient cooling on its own. Some form of water cooling is necessary. The evaporation of the water provides the cooling effect, so some air movement is still required, but not necessarily large quantities of water.

Water can be applied directly to the pigs in the form of a spray or drip system installed in the shed. Heat load in the building can be reduced by spraying water onto the roof, but water usage is high. Ducted, fan forced evaporative cooling can also be used.

Good design of such systems is necessary to minimise water usage, and reduce extra problems of corrosion, food spoilage and so on.

Heating

It is also necessary to maintain a warm environment in cold weather. Newborn piglets are particularly sensitive. Zone heating is applicable to such situations, where heat is applied to one part of the pen. Pigs needing the heat will be attracted to that area, the advantage being that not all the shed needs heating. Electricity and gas are the common energy sources, but there are many different types of heaters available. For general heating applications, radiant heaters, space heaters or under-floor heating (electrical resistance cables or hot water pipes) are in use.

Even in cold weather some ventilation is still required, which will affect heating requirements.

Insulation

Heat load in summer and heat loss in winter can be greatly influenced by shed insulation. Reflective insulation helps in both summer and winter situations, but bulk insulation is necessary for best performance in winter. A large range of insulating materials and products are available, each with their particular attributes. Consider their installation cost, thermal performance, durability in the harsh environment, attractiveness to nesting vermin, and behaviour during a fire. Some require specialist contractors to apply, and thorough preparation at the point of application.

Fibreglass batts would not be suitable unless fully enclosed by a ceiling, because they will deteriorate quickly, and be a haven for rodents. Polyurethane and polystyrene products are in wide use. Reflective foil needs to be supported by wire mesh, and is subject to damage if too loosely draped. A sealant or vapour barrier may also be required. Under-floor methods of insulation have been used, such as construction of a honeycomb type layer underneath.

Water supply

A substantial supply of good quality water is required; around 10 litres per pig per day for drinking, perhaps double that again for washing down. Drinking water supply is usually from pressurised pipes to drinking nipples or outlets. Storage and reticulation of water needs careful planning and design, particularly for multiple shed operations, and where water cooling systems are in use.

Feeding

A variety of feeding systems are available; bucket or pail by hand from a trolley, self-feeders, or automatic feeders. Savings in feed, by reducing wastage, are worth achieving, so a well designed and efficient feeding set-up is advantageous.

Special purpose buildings

Figure 9.11 *Waste management is a major factor in the planning, design and construction of piggeries*

Sufficient storage for two weeks feed should be available. On larger operations specialist advice is necessary to integrate the feeding method with the handling, milling, mixing and delivery aspects of the feeding system.

Waste management

The simplest method of waste management (but most laborious and probably least satisfactory) is where the pigs are housed on a solid floor, and it is hosed down into a drain as required. The drain delivers effluent to a collection pond or tank.

Modern systems have pigs housed on floors that are partly or fully slatted, with a drainage system below. Occasional flushing with water delivers the effluent to the collection point. Some systems have the collection pond under the slats, continuously circulating, but these cause problems of equipment maintenance, disease transmission, high humidity and odour.

The design of the waste management system is important to ensure that flushing occurs satisfactorily, without too much water. The slatted floor, drain and flushing equipment need to be designed carefully, compatible with pollution control guidelines; specialist advice is required.

Having collected the effluent, it must be stored, treated and disposed of in an appropriate manner. Disposal is generally by land application, taking advantage of its nutrient content, but being careful to avoid odours and pollution of streams and groundwater. Various methods of teatment, needed prior to land application, are available. Each requires careful consideration of capacity, throughput and biological processes. Composting methods may

Figure 9.12 *Equipment in place for feeding, watering and warming broiler chickens*

also be an option. This aspect of pig housing must be carefully considered during the planning stage, ensuring sufficient land area is available for disposal, and minimising the effects of other site constraints.

All methods will need the approval of local authorities.

Other specialised buildings

Intensive poultry housing

Like other animals, housed poultry (layers and broilers) have well defined environmental requirements for optimum performance. They generate a large amount of body heat, and the principles of selection of construction materials, natural and forced ventilation, heating, evaporative cooling and stocking density are equally relevant, but applied in a different context. Provision for specialist feeding, watering and egg collecting equipment needs to be considered.

Figure 9.13 *Caged birds for intensive egg production*

Figure 9.14 *In this rotary milking shed, cows move on and off the slowly rotating platform unassisted, and they move past the operator. Cups are automatically removed at the end of the cycle. In some milking systems, cows are electronically identified as they enter the bail, their individual feed ration is automatically delivered to them, and their milk production recorded*

Dairy buildings

Environment control is less important for milking sheds provided the milking operation is protected from the elements. However, because of the labour intensive nature of the operation, a lot of planning and design detail is applied to maximise efficiency (as measured by cows milked per operator per hour), whilst maintaining adequate attention to udder stimulation and hygiene. A herringbone layout, with the operator at a lower level than the cows to avoid stooping, can improve productivity. For larger herds rotary milking parlours are in use.

The size of the milking shed depends on the number of cows in the shed at any one time. This need not necessarily be a large number to achieve satisfactory throughput. It also depends on the way the cows are stalled (on an angle, square on to the milker, or on a rotary platform). These decisions involve a number of secondary factors, such as the time required to prepare the cow, and the time necessary to milk her out. Feeding the cows in the stalls is increasingly common.

Other factors influence aspects of the shed design; the number of milking units, feeding method, the direction the cows enter and exit the shed, the preferred location for bulk milk storage, and details associated with the size and elevation of the pit.

Figure 9.15 *Note the sunken pit for the operator in this herringbone milking shed. Good selection of materials and surface finishes is required, as is an overall design compatible with the installation of the necessary equipment*

The selection of materials, and their strength and durability, is important. Concrete must have quality and surface characteristics to sustain heavy traffic and high pressure hosing down. Wall surfaces need to be finished in a way that is easy to clean. Gates and rails need to be sturdy and attached securely to handle the frequency and pressure of cow movements.

Drainage of the wash-down waters from the building, and drainage from the site generally, must be accommodated. Run-off and wash-down water must be collected, and disposed of by land application to prevent local pollution. Check with local authorities for specific requirements.

The milk storage vat is housed in its own room adjoining (but not opening on to) the milking shed. Hygiene and temperature control are important, so materials selected for their ease of cleaning and insulation properties need to be considered. Separate housing again for vacuum and refrigeration equipment will be required.

Stables and stalls

Individual stalls can be easily fitted to conventional building designs. For large animals such as horses and cattle stall size of 3.6 x 3.6 metres is typical. A minimum building height of 3.0 metres is recommended, but this allows little room above the animal for effective ventilation, unless eave vents are combined with ridge vents. Even so, this risks stagnant air low down in the stall; ventilation requirements are just as important here, particularly since

Special purpose buildings

insulation is not commonly used. Stall frames need to be strong, and adequately secured against animal pressure, kicking and so on. Adequate lighting is required.

Materials should be selected for resistance to corrosion, but also for their durability around large animals. The bottom half of the stall is frequently of timber (solid rails or heavy duty plywood) but capped with steel to prevent chewing. Wood tends to better absorb kicking, provided it is sufficiently strong that it doesn't break. It may be an advantage to have the stall designed so broken timbers can easily be replaced. Old conveyor belt can also be useful to absorb kicks. The top half of side panels could be timber as well, but with a strong mesh or grill to the front to guard against chewing. Doors need to be hinged securely, and provide bruise free openings of sufficient width.

Floor materials are often laid brickwork, covered with straw or sawdust. Special purpose flooring bricks are available. The floor should be easy to clean. Wall panels in contact with the floor risk accelerated deterioration, particularly where moisture from urine, leakage from drinkers, or rain, is present.

Some operators include an exercise yard adjacent to each stall, with an external gateway in the building wall.

Consider the flammability of all materials, and the means of exit in the event of fire.

Figure 9.16 *The stalls in this shed accommodate stud cattle*

References

Australian Standard AS 2867–1986, *Farm Structures — General Requirements for Structural Design*. Standards Association of Australia, Sydney, 1986.

BARBER, A.A. & FREEMAN, R.B. *Design of Shearing Sheds and Sheepyards*. Inkata, Melbourne, 1986.

CONNELLAN, G. "Greenhouse Structures — Coverings and Environments". *Australian Horticulture*, September 1982

DAWKINS, D. & CUSACK, D. (eds). *Australian Domestic Construction Manual, Part 2*, NSW edn. Standards Australia, 1993.

LYSAGHT BUILDING INDUSTRIES. *Steel Roofing and Wall Installation Manual*. John Lysaght (Australia) Ltd, Sydney, 1991.

REDDING, G. *Functional Design Handbook for Australian Farm Buildings*. University of Melbourne, Melbourne, 1981.

SOUTHORN, N.J. & BALDWIN, B. *Property Planning and Development*. External Student Study Guide, Orange Agricultural College–The University of Sydney, 1995.

TAYLOR, G., KRUGER, I. & FERRIER, M. "Plan It — Build It". *Australian Pig Housing*, Series No. 2, NSW Agriculture, 1994.

Acknowledgments

For making illustrations and technical material available, the author and the publisher are grateful to:

BHP Building Products Pty Ltd
Tamarang Enterprises PtyLtd
Econo Supershed Pty Ltd
Ramset Fasteners (Australia) Pty Ltd
Gang-Nail (Australia) Limited
W A Deutscher Pty Ltd
NSW Workcover Authority

To Joanne for typing the manuscript
Brett Upjohn for assistance with some of the illustrations
Mal Lukins for assistance with some of the original material
and Paul Conti for reviewing part of the manuscript.

Index

A
abattoirs 22
access 14, 117–118, 133, 139
amenities 13, 136
angle 52, 60
ant cap 100
arch 95, 96
aspect 17–18
Australian Height Datum (AHD) 60, 61

B
base plate 86, 91
bay 18, 81
beam 77, 81, 82, 89, 90, 92, 98
bearers 99, 100, 101
belling 86
borers 44–45
bracing 71, 84, 85, 93
–apex 81, 82
–cross 81, 82, 83, 84–85, 89, 92, 124
–knee 81, 82, 89
brick 100
Building Application 22
bunyip level 63

C
capping 110, 111, 125
cement 28, 29
centralisation 13
chain 56
channel 28, 77
Chemset® 51
cladding 50, 73, 74, 84, 88, 89, 92, 97, 102–114, 123, 128, 129
Code of Practice—Safe Work on Roofs 10
cold galvanising 27
Colourbond® 102, 103
columns 18, 28, 43, 74, 79, 82, 86, 89, 90, 92, 93, 100, 122, 124
compass 60
compression 24
concrete 12, 28–38, 97–99, 113, 124, 138
–aggregate 28, 30, 33
–bleeding 30, 36
–construction joints 35
–curing 31, 37
–finish 34, 36, 37
–hydration 29, 37
–plastic 28, 29–30, 35
–reinforced 12, 33, 37, 100

contractors 23
cooling 117, 130, 139
see also ventilation
coolstore 19
corrosion 21, 27, 34, 121, 122,–123, 127, 133, 138, 145
Custom Orb® 103

D
dairies 117, 143–144
density 44
Development Application 22
doors 89, 90, 102, 109, 145
downpipes 110, 117, 122
drainage 15, 16, 17, 21, 87, 110, 111, 116, 124
ductility 24
durability 44
Dynabolts® 50, 51

E
eaves 17, 21
effluent disposal 20–21, 117, 141
electricity 13, 115–116
electronic distance measurement 57
elevation 52, 60
extrusion 27

F
fasteners 46–51, 105–107
–bolts 46, 77, 82, 90
–chemical adhesive fasteners 50
–framing anchors 47
–joist hangers 47
–nailing plates 47
–nails 50, 84, 105
–washers 105
feedlots 22, 140
fire hazards 13, 21, 45
foil 19
footings 11, 12, 15, 16, 19, 23, 80, 85–86, 87, 89, 91, 95, 99, 100, 122, 127, 138
forging 26
foundation 11, 12, 16, 32, 85, 98, 100
fuel 13

G
galvanising 28, 102,–104, 122, 127, 138
girts 50, 80, 82, 84, 89, 90, 92, 93, 100, 103
greenhouses 18, 19, 83, 96, 97, 117, 118, 126–130

gussets 81, 83, 89
gutters 21, 89, 91, 110, 116, 122
–box 81

H
hardwood 38, 76, 99
hay sheds 13, 92, 96
Health Department 117
heartwood 38–39
heating 116, 130–131, 140
holding down bolts 87, 100

I
inspections 120–121
insulation 19, 43, 108, 140, 145
insurance 23

J
joists 99, 100, 101

K
kit sheds 17, 22, 78, 89, 91
Klip-Lok Hi-Ten 103, 104

L
ladders 9, 10, 138
license 23
lifting 9
lighting 115, 116, 134, 135
load 24, 71–76, 95, 98, 100, 101, 137
–duration of 39, 42
–estimation 71
–wind 72, 90, 127
Loading Code 72, 75
local terrain 18, 75

M
machinery sheds 17, 83
maintenance 13, 14, 120–125

N
noise 43

O
offset site survey 57, 58, 59
optical distance measurement 56
optical instruments 60, 61–63

P
pegging 66–70
personal protection 9
piers 12, 17, 98
piggery 18, 19, 20, 22, 97, 99, 103, 121, 138–142
pipe 28, 77, 92
planning 11–23
plans 11, 22, 52–70
–scale 53

pole-in-ground 92
portal frame 80–94, 124
poultry housing 20, 119, 142
preservation 27–28, 44–45, 93
proximity 13
purlin 28, 43, 50, 80, 81, 82, 83, 84, 89, 90, 92, 93, 94, 103, 104, 123
C-section purlins 84
Z-section purlins 84

R
radius 57, 58
raised floor 17, 72, 99–101
rectangular hollow section (RHS) 28, 77
refrigeration 118
see also coolstore, cooling, ventilation
regulation, local council 21
roads *see* access
rodents 113
roof 10, 13, 19, 72, 73, 74, 81, 82, 92, 110–111
run-off *see* drainage

S
safety 9–10, 15, 138
scaffolding 9
shearing shed 17, 18, 20, 43, 72, 90, 99, 100, 101, 108, 120, 121, 131–137
silicone 110
silos 137–138
site 11, 22
skylights 19
softwood 38, 76
span 18, 76, 78, 80, 81, 82, 102
Spandek Hi-Ten 103
specifications 11, 22
split level 17
stables 144
steel 24–28, 47, 76, 83, 92, 100, 102, 138
storage 13, 17, 72, 99, 121, 136
strain 26
stress 26, 76, 77, 78
subsidence 12, 124
suspended loads 19

T
tape, measuring 56
tension 24, 25, 76
termites 44, 112, 120
timber 38–45, 48, 76, 84, 92, 93, 99, 100, 101, 112, 122, 124–125, 145
–defects 40
–grain 39, 40
–moisture 39, 41
–shrinkage 41, 44
–species 40, 42
–strength 39, 41, 42
–stress 43
–temperature 39, 42

tracheometry *see* optical distance measurement
traffic 118
transport 13
triangulation 54, 57, 59
Trimdek Hi-Ten 103
Trubolt® 51
truss 92, 94, 95, 123

U
U-bolt 86
universal beam (UB) 28, 77

V
ventilation 18, 19, 116, 118–119, 127, 128, 129–130, 134, 135, 139
–cross-flow 19, 96, 97, 118, 129, 139
–forced 97, 128, 139
–ridge 19, 97, 118, 119, 128, 129, 139
vibrators 36

W
waste disposal *see* effluent disposal
weatherproofing 89, 109, 110–111
welding 47, 78, 77, 122
wheel, measuring 55–56
wind 17, 75–76, 85
–tunnel 18
see also load, wind
windows 88, 89, 90, 91, 102, 109
Workcover 10, 23
workers compensation *see* Workcover

Y
yield stress 24, 26

Z
Zincalume® 102, 103
–screws 105